A FREE GIFT THAT MAY BE OVER UNITY OR FREE ENERGY TO THE WORLD

ISBN:
ISBN-13:

DEDICATION

This book is dedicated to Jesus Christ who has had a large influence in my life on how to conduct one's life. I am thankful for the forgiveness of sins and the promises He has given me about the new life I now have in Him. My wife and I wrote the book "Please pass the PEW" in love for Jesus Christ.

CONTENTS

Mechanical make up of Proto-type motor one
Several drawings of proto-type motor
Multi-purpose rotor assembly
     Old work for most people is new work, so here it is
Flow Through Electro-Magnetic Motor: Theory of Operation:
Mechanical Characteristics:
Figures 1000 – 1003
Multiple Magnetic Flow Through Motors:
Permanent Magnet Ring;
Electro-magnet assemblies;
Bearing Assemblies:
Multiple Flow Through Electro-Mechanical Motor Assemblies;
Flow Through Motor Angles
Motor Addition Design;
Power circuits;
Cooling Coils and Fin Assemblies;
Vehicle Applications:
Flying "SKY CRAFT"
Combination sky and sea craft;
Object Propulsion Flow Through Motor;
Generator and Breaking Systems;
Motor Start Up Options;
Final Conclusions;
Reason for giving the technology away free to the world

INTRODUCTION:

This is the first of 5 "**free energy**" books I have written. All of my books are all in theory mode at this time because I do not have third party verification of my motor and generator designs.

This book, "A Free Gift That May Be Free Energy or Over Unity to The World", is my first book that includes thoughts of motor designs started in 1969. This book includes more than three theories that should improve electric motor performance. The three main theories are "Flow Through Motor Technology", "Three Layer Electro-Mechanical Movement Technology", and "Recovery from Collapsing Magnetic Field Technology". The paper back copy of this book is black and white in order to keep the cost down for more people to be able to purchase the book.

The second book "Permanent Magnet Torque Harvesting" repeats some of the work shown in this book of the "Three Layer Electro-Mechanical Movement Technology" but then takes the technology and shows several applications of it in color. This book also shows generator designs with a new technology I call "Indirect Power Generation".

The third book "AAAAA Amazing Apparatuses Applying Abundant Abilities", has both motor and generator designs with a new technology I call "Chain Reaction Flux Switching Technology". This book is in color. This technology being in theory, has great potential if it is proven to be viable. These motor and generator designs are a great place to start with to use your creativity to come up with your own motor and generator designs.

The fourth book "Motionless Switching Magnetosphere Electric Generator", takes the "Chain Reaction Flux Switching Technology" to the next level for electrical energy. There are several generator designs in this book that are looked at in this book. The generator would change the world if the technology can be proven by a third party and then promoted.

The fifth book, "The Core to Free Energy", is the last book I have written that uses two new technologies not used in my other books. Many of the designs in this new book use a combination of the old technologies along with the new technologies I have come up with for the motor designs. Some of the mechanical designs are totally unique from the mechanical designs from the first four books. I believe these are the best motor designs of all of the motor designs I have written or seen. If you are to purchase any of the five technical books, then this is the book to buy.

In this book, I believe I have invented a motor that meets this criterion called free energy that I want to freely share with the rest of the world. I have worked on many different motor designs since 1969. I have kept them private over the years. I could be rich in royalties by patenting them but I want to freely give them to the world so that the world can be a better place for all of us. I have done the work on my own and to my knowledge have not copied other people's work. I do not know if someone else has not come up with the same motor designs. With over 50 million patents in the world, I could never claim that no one else has come up with the same conclusion. I want to share what I have come up with on my own so that other people can build these motors to save on their energy expenses. Not only would these motors save people money, but also on the dependence on fossil fuels.

I am a visual learning person. I like visual illustrations in explaining things to me, so I will do the same for you in this book. Several sketches and drawings are used to give you a better mental picture of what I am writing about with the different new technologies I will be sharing with you in this book.

The next chapter gives you the best motor design to start with and the following chapters that show how I worked up to this design and going into greater detail how these technologies came about with several applications of how these technologies can be used in design work.

The designs are something that most hobbyist can accomplish building with moderate expenses for tooling and materials. Now some designs go into more detail than others about the details of how to build them. But I try to give people enough information so that they will be able to fill in the blanks of the assembly of the devices I have in this book.

I do have some designs that take more money than most people can afford to pay for, but I wanted to include where these new technologies can go down the road in the future for companies to build these products.

5 TO 1 PM TO EM TORQUE DISK MOTOR:

The following drawing of the most efficient motor I have designed to date using the "Three-Layer Electro-Mechanical Movement Technology".

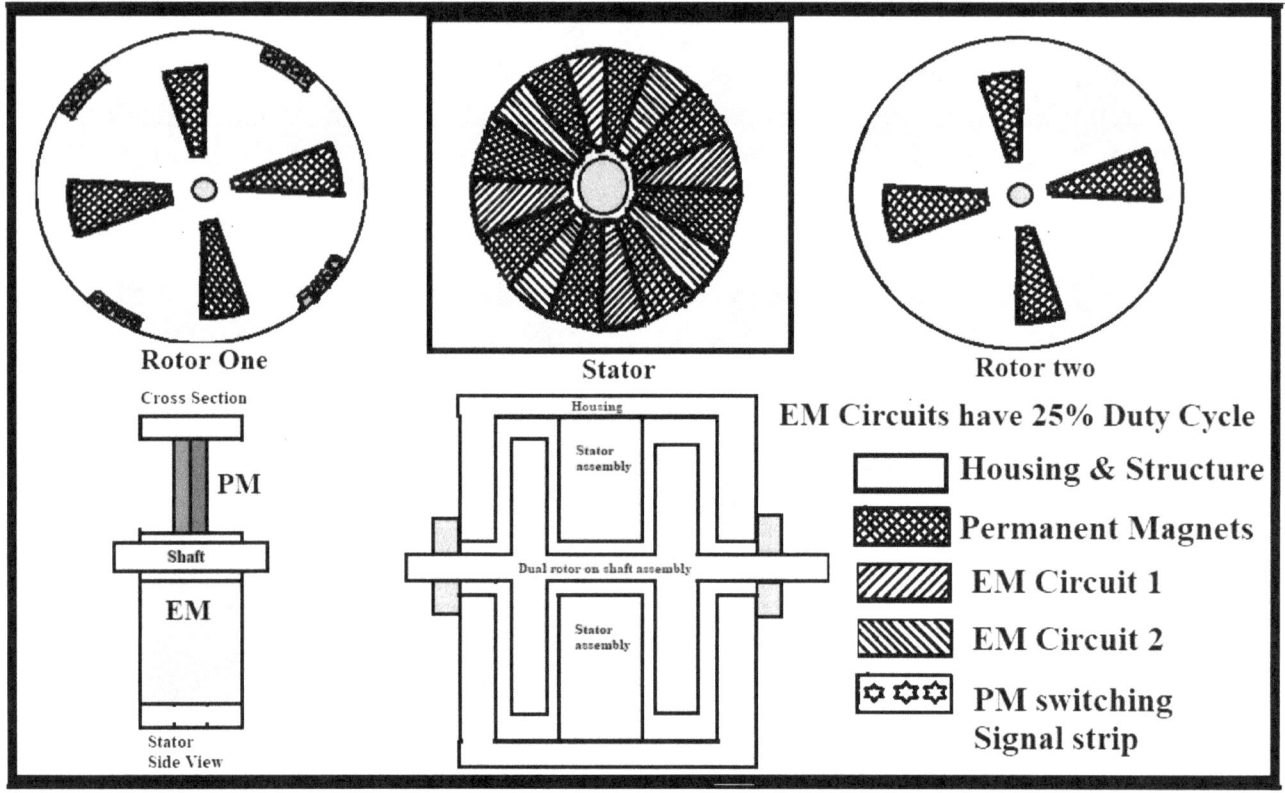

This chapter has the latest development of a motor using the "Three-Layer Electro-mechanical Movement Technology" that I have developed. If you are not familiar with the technology then you may want to read this chapter last.

The "Three-Layer Electro-Mechanical Movement Technology" creates a new functional structure of the stator assembly where the interaction between the rotor and stator assembly is changes from one of repulsion to one of attraction. This functional structure is turned on and off again and again in such a way that the motor will have forward torque in it with the power duty cycle set to a level of 25% for each of the two electro-magnetic circuits. I will go into details of how this motor operates with the torque from 5 permanent magnets while the activation of only one electro-magnet in other chapters.

You only have to pay for the torque from one electro-magnet while utilizing the torque from five permanent magnets when using this new technology.

The core in the electro-magnet does not need to be as thick as the permanent magnet while the coil can be thicker than the permanent magnet. With this configuration, forward torque can remain in the function of the motor while the core of the electromagnet will provide reusable power for the motors power circuit. The following drawing is an easy to build circuit for the hobbyist to build. This circuit captures electrical energy from the collapsing magnetic field from the core of the electro-magnet when the power is turned off, to be used in the following power cycle of the electromagnet circuit that is operating at a duty cycle of 25%.

The next drawing is one optional electrical circuit that can be used to operate the motor above. It is good to incorporate as many technologies as possible to improve the performance at the system level of free energy devices. The second drawing is the writing of the function of this next drawing. I am doing this with several drawings in order to make the information more viewable for you.

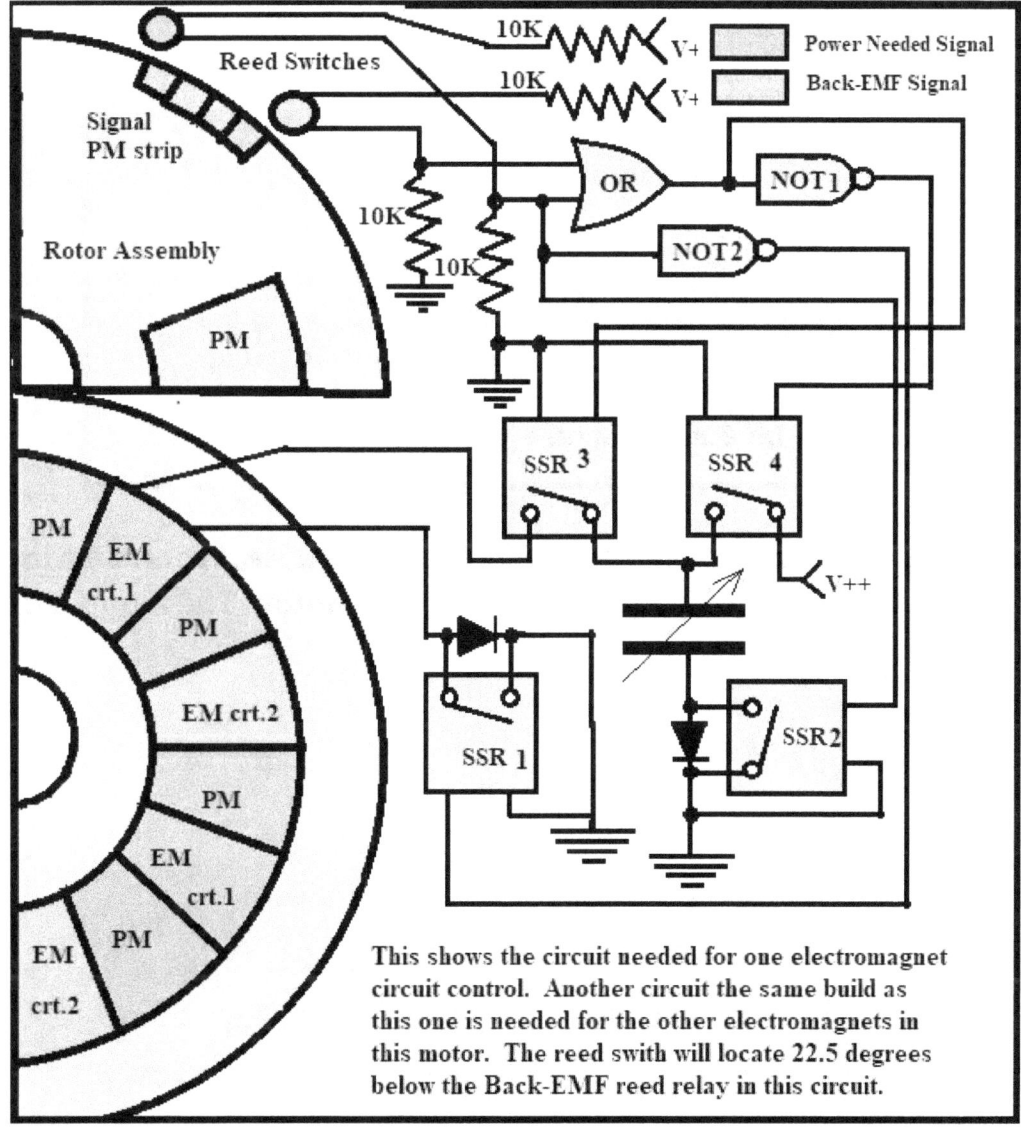

This shows the circuit needed for one electromagnet circuit control. Another circuit the same build as this one is needed for the other electromagnets in this motor. The reed swith will locate 22.5 degrees below the Back-EMF reed relay in this circuit.

Defining logic in tank circuits used in three layer disk motor

The motor is devided into 16 segments of travel each having 22.5 degrees to thier travel.

The stator has two electromagnet circuits, one to operate in the first segment of travel and the other one to operate in the third segment of travel. During the second and forth segments of travel, the air core electromagnets are turned off and the permanent magnet in the rotor assembly interact with the permanent magnets in the stator assemblies to rotate the rotor 22.5 degrees.

So the following logic occures in the circuit as follows:

Segment 1: The Power Needed Signal reed switch closes creating a posative signal on its signal line

A. The signal goes through NOT2 being a low signal. This opens the switch in SSR1 restricting the current flow in one dirrection. As the current flows from the Tank capacitor, we do not want back EMF to occur during this time.

B. The signal is feed to an OR gate. The OR gate will create a posative signal that lasts through segments 1 and 2. This will tie the electromagnet and capacitor together as a Tank circuits through these to segments of travel using SSR3.

C. The OR gate also feeds NOT1 gate that goes to SSR4 to make sure the the power supply is not connected to the Tank capacitor at this time.

Note: SSR2 stays in the same condition That of shorting the diode so current flows into the coil.

Segment 2: The Back-EMF signal closes creating a posative signal on its signal line.

A. This Goes to the OR gate which feeds SSR3 to keep the Tank coil and capacitore connected so the back-EMF can be collected into the capacitor.

B. The signal goes to the SSR2 relay to open the swith activating the function of the diode. This allows charging only for the capacitor in the Tank circuit.

C. Since the "Power Needed Signal is now off, SSR1 is shorting out the diode at the bottom of the Tank electromagnet allowing the current to flow into the capacitor.

Segment 3 and 4 are the same:

SSR1 Through NOT2 causes switch to close shorting diode.

SSR2 is open so the dide at the bottom of the Tank circuit is active.

SSR3 has a low signal from the OR gate causing the electromagnet circuit to be open.

SSR4 comes from a NOT OR logic causing the swich to be closed in order to top off the charge of the Tank capacitor.

Jay Lunke Sept. 30, 2020

The disk motor design can easily be expanded into several layers. These layers add a lot of power to the motors, creating a lot of power per square inch of motor. The number of applications of this motor are limitless.

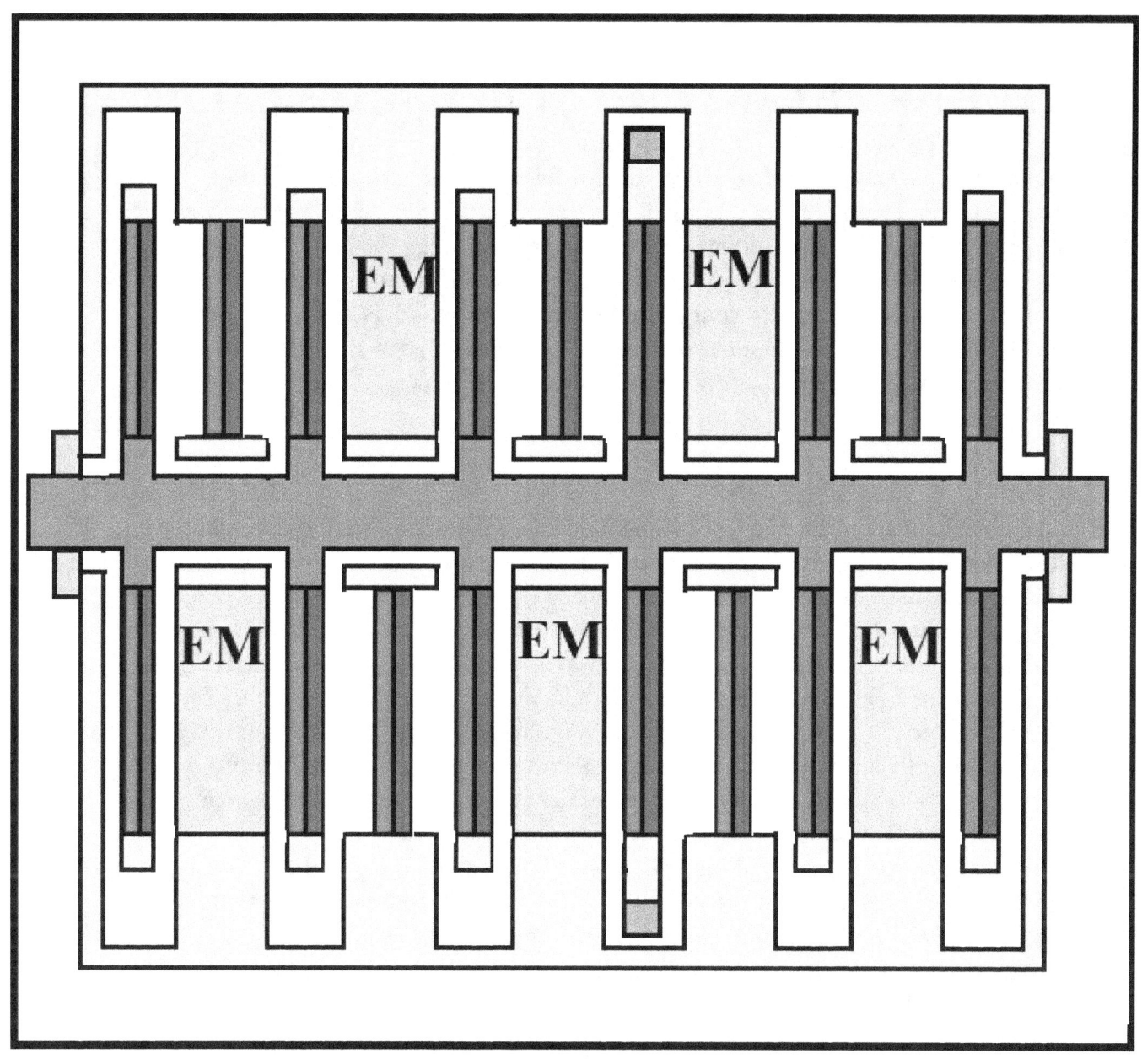

Milti-Disk Five to One PM to EM Disk Motor

The multi-disk format of the 5 to 1 PM to EM disk motor has the potential of moving motor technology from wagons being pulled by square wheels to using round wheels when power by the modified tank circuit with steering diodes.

It one one thing to have a motor that is overunity, but if it is bulky and is not much over unity, then does not make sence to use it in a vehicle. This motor design with the power circuit has more than enough power in torque to power an efficient generator to power the motor at the same time have enough power left over to have its shaft power the drive train of the vehicle.

If and only this is true about this new technology

Being an old man soon to be living on social security, I will not be able to build and advance this technolgy. It is up to people like you who can see the potential of this technology and then to build prototypes to prove its potential for the world. I have to admit that when I look at other peoples motor designs, it is hard for me to have that desire to build thier design because I always want to focus on my own designs. I reached out for help and I was asked to pay 18 times of what I was making for that help. I have not accepted money in the past and I am not seaking money in the future. What I am asking for is people to examine all the postings I have made at this site that includes much more details of this new technology which includes the modified tank circuit with steering diodes.

Just think of the good it would do for the average person of the world to have a motor that could provide all of the power they needed in a self contained system without having to buy fuel or electrical energy to operate it.

By building and posting the results of this new motor technology would be great. If the results are that the technology does not meet the OU status, I could live with that. Even if the results was a more efficient motor using less electrical energy per mile traveled in your car would be a big plus for everyone.

Jay Lunke Oct. 14, 2020

The following four drawings show four electromechanical movements that are involved in the "Three Layer Electro-Mechanical Movement" technology. There is a more in-depth definition in later chapters in this book.

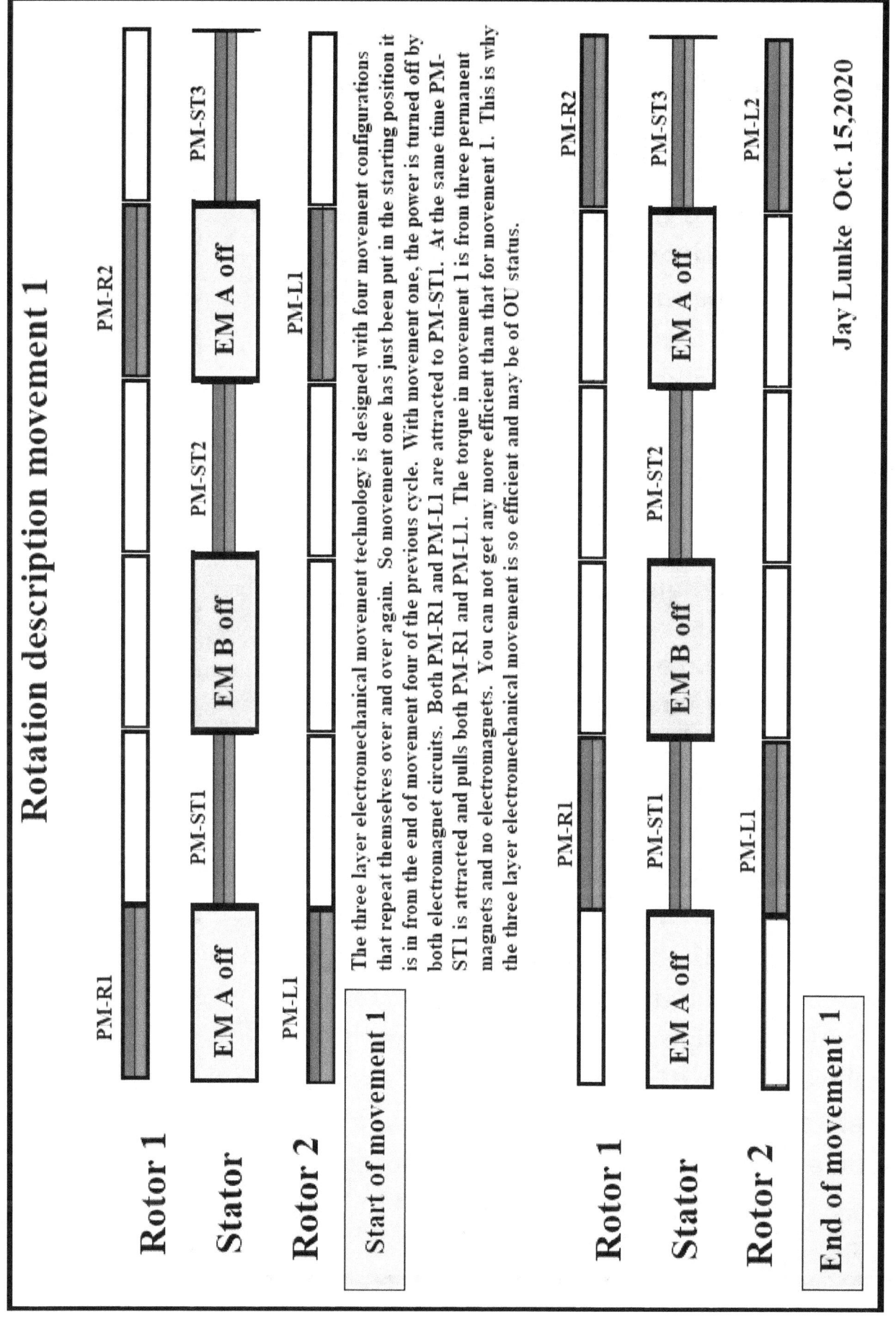

Rotation description movement 1

The three layer electromechanical movement technology is designed with four movement configurations that repeat themselves over and over again. So movement one has just been put in the starting position it is in from the end of movement four of the previous cycle. With movement one, the power is turned off by both electromagnet circuits. Both PM-R1 and PM-L1 are attracted to PM-ST1. At the same time PM-ST1 is attracted and pulls both PM-R1 and PM-L1. The torque in movement 1 is from three permanent magnets and no electromagnets. You can not get any more efficient than that for movement 1. This is why the three layer electromechanical movement is so efficient and may be of OU status.

Jay Lunke Oct. 15,2020

This is only the first of four movements that is designed into the three layer electromechanical movement. All four movements are needed for this technology to function. This technology can only function with one rotor permanent magnet for every four possible positions on the rotor. This is because this technology creates a flux wave that continually travels around the stator assembly by continual reconfiguration of the stator assembly. The rotor permanent magnet rides on the crest of this wave through the changing torques between the rotor and stator assembly permanent magnets.

When you conpare the torque from electric and magnetic motors on the market today, I have not seen any that have a torque of five to one or more. There have been some designs that have more magnets to electromagnets, but some of them have multiple magnets to create one torque from the collection of them. In those designs, the motors can become large in size for the amount of torque you get from them. This design offers a lot of torque for a small package with minimal required external power to produce that torque. This should produce many applications for this motor design. Now there are restrictions of how small it can be made.

Rotation description movement 2

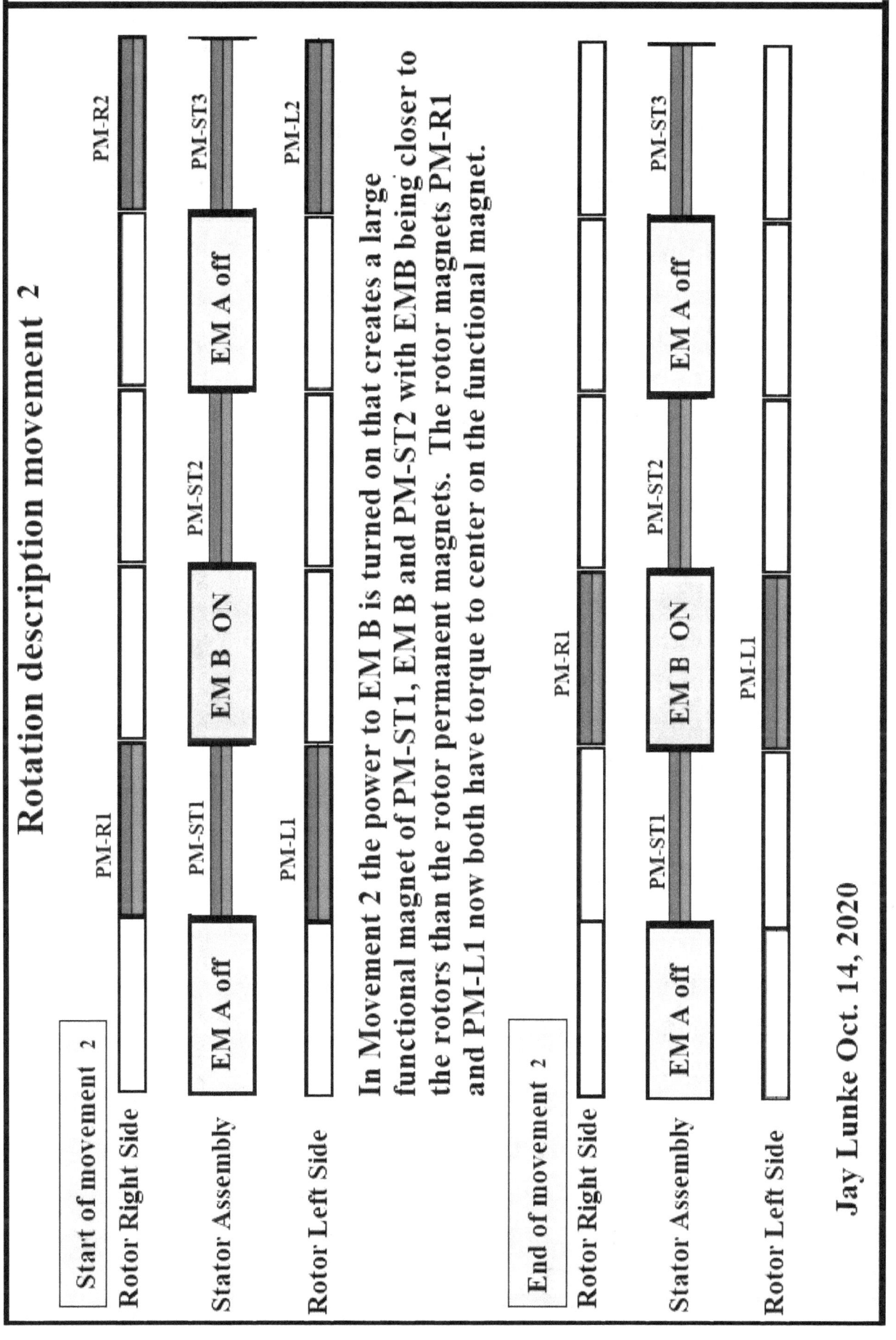

Start of movement 2

Rotor Right Side — PM-R2 / PM-R1

Stator Assembly — PM-ST3 (EM A off) / PM-ST2 (EM B ON) / PM-ST1 (EM A off)

Rotor Left Side — PM-L2 / PM-L1

In Movement 2 the power to EM B is turned on that creates a large functional magnet of PM-ST1, EM B and PM-ST2 with EMB being closer to the rotors than the rotor permanent magnets. The rotor magnets PM-R1 and PM-L1 now both have torque to center on the functional magnet.

End of movement 2

Rotor Right Side — PM-ST3 / PM-R1

Stator Assembly — PM-ST3 (EM A off) / PM-ST2 (EM B ON) / PM-ST1 (EM A off)

Rotor Left Side — PM-L1

Jay Lunke Oct. 14, 2020

This is a unique function of the three layer electromechanical movement is to reconfigure the stator function by changing the functional configuration of the stator magnets several times during the operation of the motor using as many permnent magnets during the reconfiguration in order to reduce the number of electrical magnets needed in the total amount of torque produced in the motor. During movement 2, PM-R1 has torque to align with EM-B, PM-R2 has torque to align with EM-B. EM-B has torque to pull both PM-R1 and PM-L1 into alignment with itself. When you combine this with the three torque of Movement 1, you end up with a torque ratio of 5 PMs to 1 EM magnet.

Out of the four movements this is the only time that power is applied to the EM-B circuit. This means a duty cycle of 25%. The EM-A circuit is off during movement 2 which allows the functional magnet to be created in the stator assembly. Now most conventional electric motors do not have permanent magnets. Magnetic motors have a 1 to 1 permanent magnet to electromagnet ratio, so have a ratio of 5 to 1 puts this motor way more efficient than other motors.

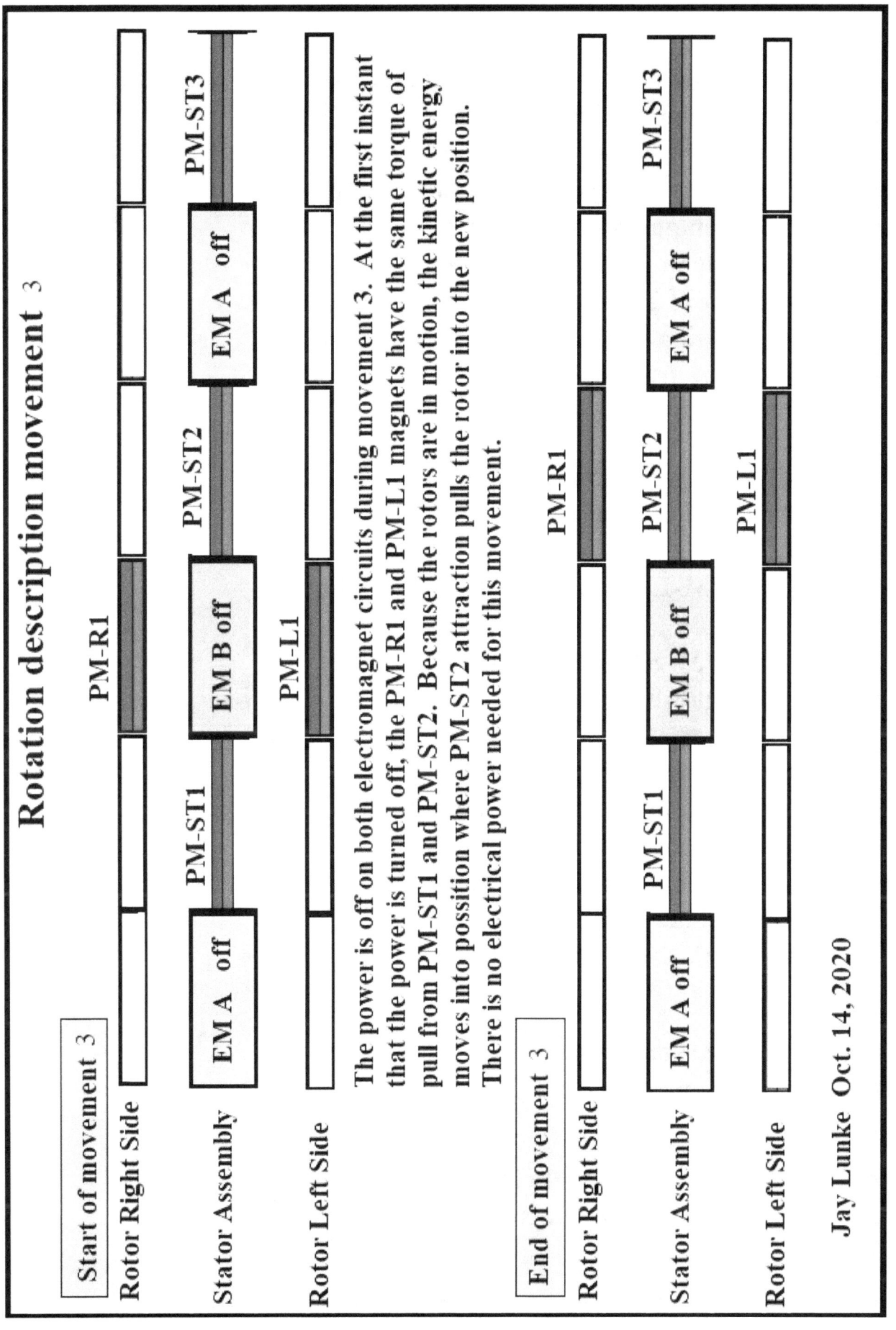

Rotation description movement 3

Start of movement 3

Rotor Right Side

PM-R1

Stator Assembly — EM A off — PM-ST1 — EM B off — PM-ST2 — EM A off — PM-ST3

Rotor Left Side

PM-L1

The power is off on both electromagnet circuits during movement 3. At the first instant that the power is turned off, the PM-R1 and PM-L1 magnets have the same torque of pull from PM-ST1 and PM-ST2. Because the rotors are in motion, the kinetic energy moves into possition where PM-ST2 attraction pulls the rotor into the new position. There is no electrical power needed for this movement.

End of movement 3

Rotor Right Side

PM-R1

Stator Assembly — EM A off — PM-ST1 — EM B off — PM-ST2 — EM A off — PM-ST3

Rotor Left Side

PM-L1

Jay Lunke Oct. 14, 2020

The three layer electromechanical movement has two of the four movements that operate without any electrical power because when the power of the electromagnets is turned off, then the torque of the permanent magnets interacting with the permanent magnets in the rotor assemblies produce movement of 22.5 degrees to occur. The duty cycle of of having only the power off to the magnets with torque from the permanent magnets creating the rotor movement is 50%.

There are three torques in movement 3.
1.) Torque from PM-R1 pulling inself to align with PM-ST2.
2.) Torque from PM-L1 pulling itself to align with PM-ST2
3.) Torque from PM-ST2 pulling on both PM-R1 and PM-L1 to align with itself.

Having two rotors with only permanent magnets on it providing forward torque in the motor doubles the power of the motor of the motor through the doubling of torque on the motor shaft. With the second rotor, the power cycles to each of the electromagnet circuits stay the same.

Question: At what torque ratio of permanent magnets to electromagnets will produce overunity.

Question: How much more efficient is this motor using a tank circuit with steering diodes.

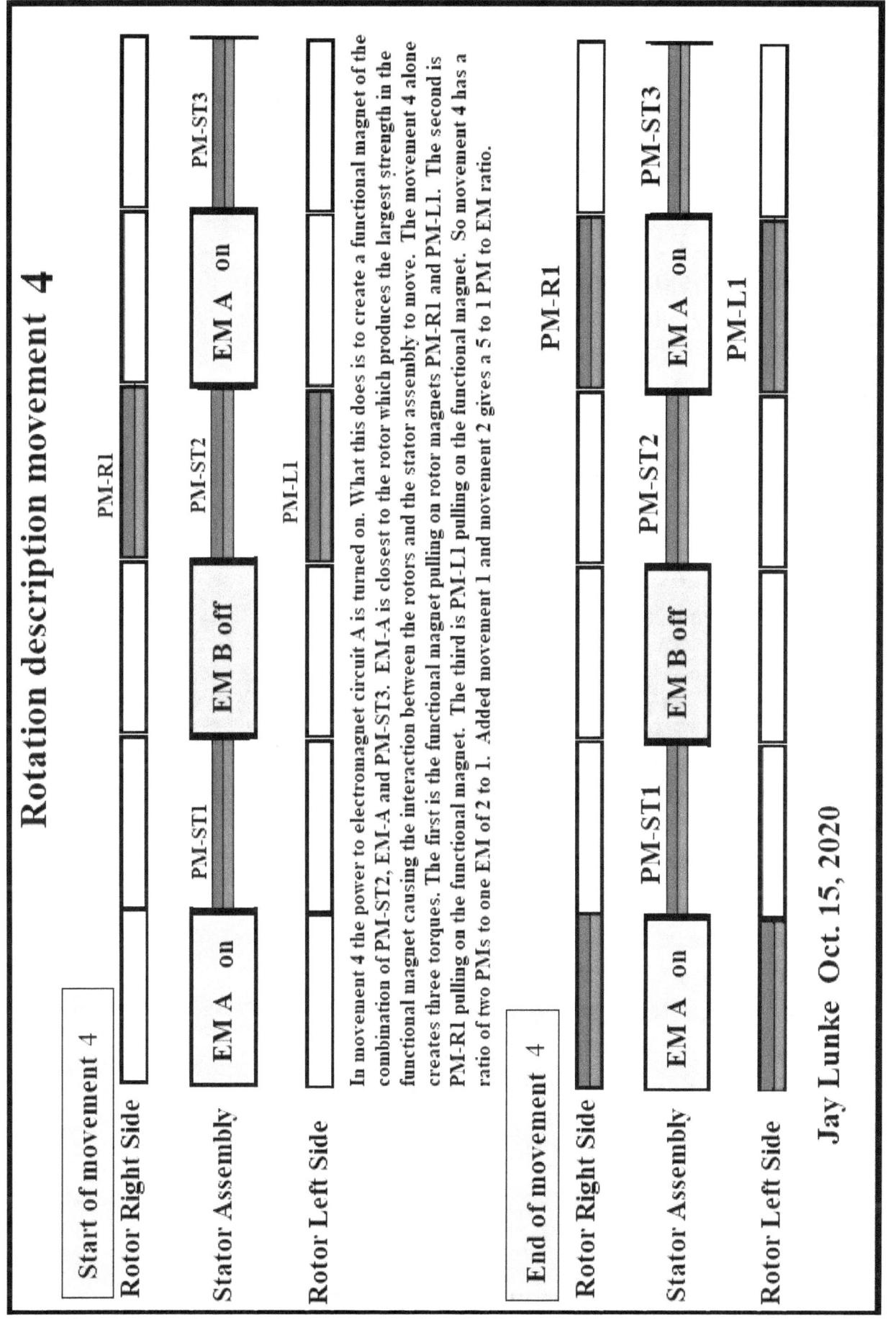

Rotation description movement 4

Start of movement 4

Rotor Right Side

Stator Assembly — EM A on — PM-ST1 — EM B off — PM-ST2 — EM A on — PM-ST3

PM-R1

Rotor Left Side

PM-L1

In movement 4 the power to electromagnet circuit A is turned on. What this does is to create a functional magnet of the combination of PM-ST2, EM-A and PM-ST3. EM-A is closest to the rotor which produces the largest strength in the functional magnet causing the interaction between the rotors and the stator assembly to move. The movement 4 alone creates three torques. The first is the functional magnet pulling on rotor magnets PM-R1 and PM-L1. The second is PM-R1 pulling on the functional magnet. The third is PM-L1 pulling on the functional magnet. So movement 4 has a ratio of two PMs to one EM of 2 to 1. Added movement 1 and movement 2 gives a 5 to 1 PM to EM ratio.

End of movement 4

Rotor Right Side — PM-R1

Stator Assembly — EM A on — PM-ST1 — EM B off — PM-ST2 — EM A on — PM-ST3

Rotor Left Side — PM-L1

Jay Lunke Oct. 15, 2020

With one stator and one rotor, the PM to EM torque ratio would be three to one. By adding the second rotor changes that ratio to five to one.

Since the power to EM-A only occures in one of four movements means that the electromagnet circuit has a duty cycle of 25%. Because the electromagnet circuits have only a 25% duty cycle in them means that they are good circuits to be powered by the modified tank circuit with steering diodes. The circuit would power the electromagnet circuit for 25% of the time followed by recovering the back EMF into the capacitor of trhe tank circuit for 25% of the time. This would be followed by topping of the tank capacitor for 50% of the remaining time before the cycle repeats itself again.

Now the motor configuration is compact allowing it to be used in many applications. When you add the fact that this design can be stacked like pancake's adds a lot of power per square inch of the motor assembly. This increases its applications a lot. If, I believe when, these motors are built and operated with the modified tank circuits using steering diodes, that this system will be meeting OU status to replace many current systems to help reduce the need for fossil fuels. Of course this is not the only OU option in the works.

Now let's take a step back a minute. The motor drawings we have looked at so far, assume a short distance difference between the stator permanent magnet and the rotor magnet compared to the distance between the electromagnet and rotor permanent magnet will not overwhelm rotor movement. The proper way that these two distances need to be in order to optimize the operation of the motor is to perform testing to determine these positions. One way to do this is to build a fixture prototype motor. With this prototype, these distances can be changed and tested several times with the results graphed out. The following prototype motor should be built and tested first before the 5:1 ratio motor to optimize the stator build.

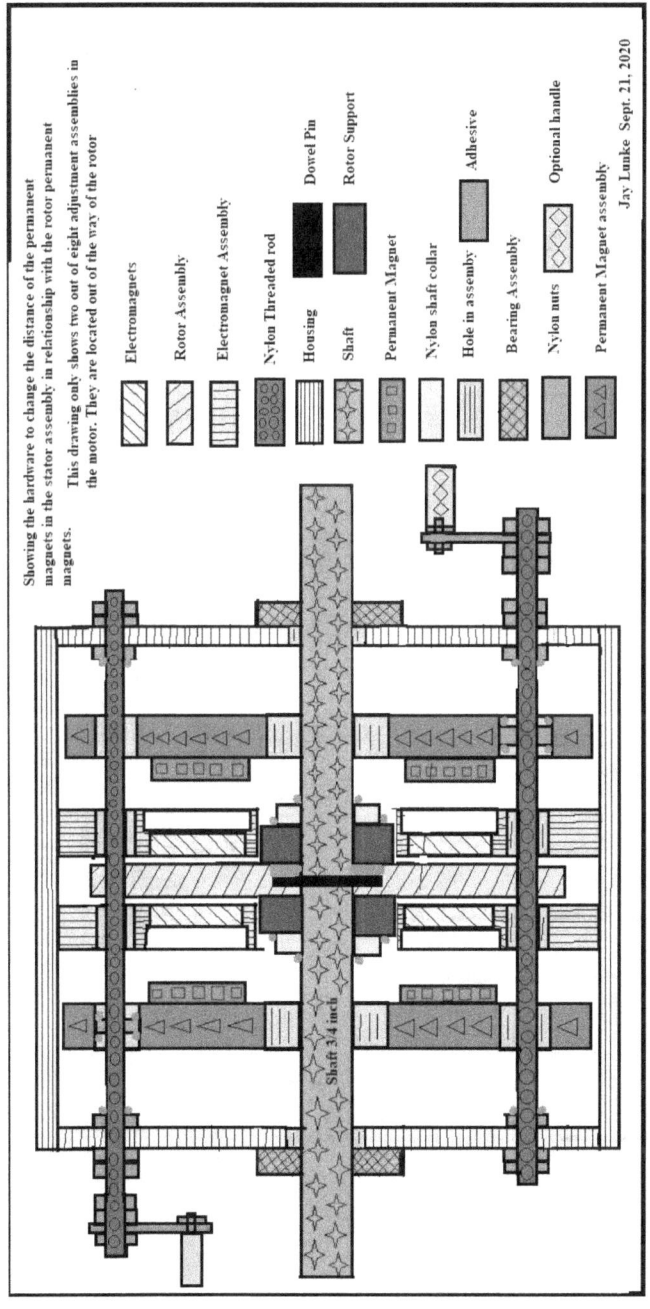

The drawing is the final prototype build with all of the sub-assemblies assembled.
The followings six drawings show more details of the sub-assemblies.

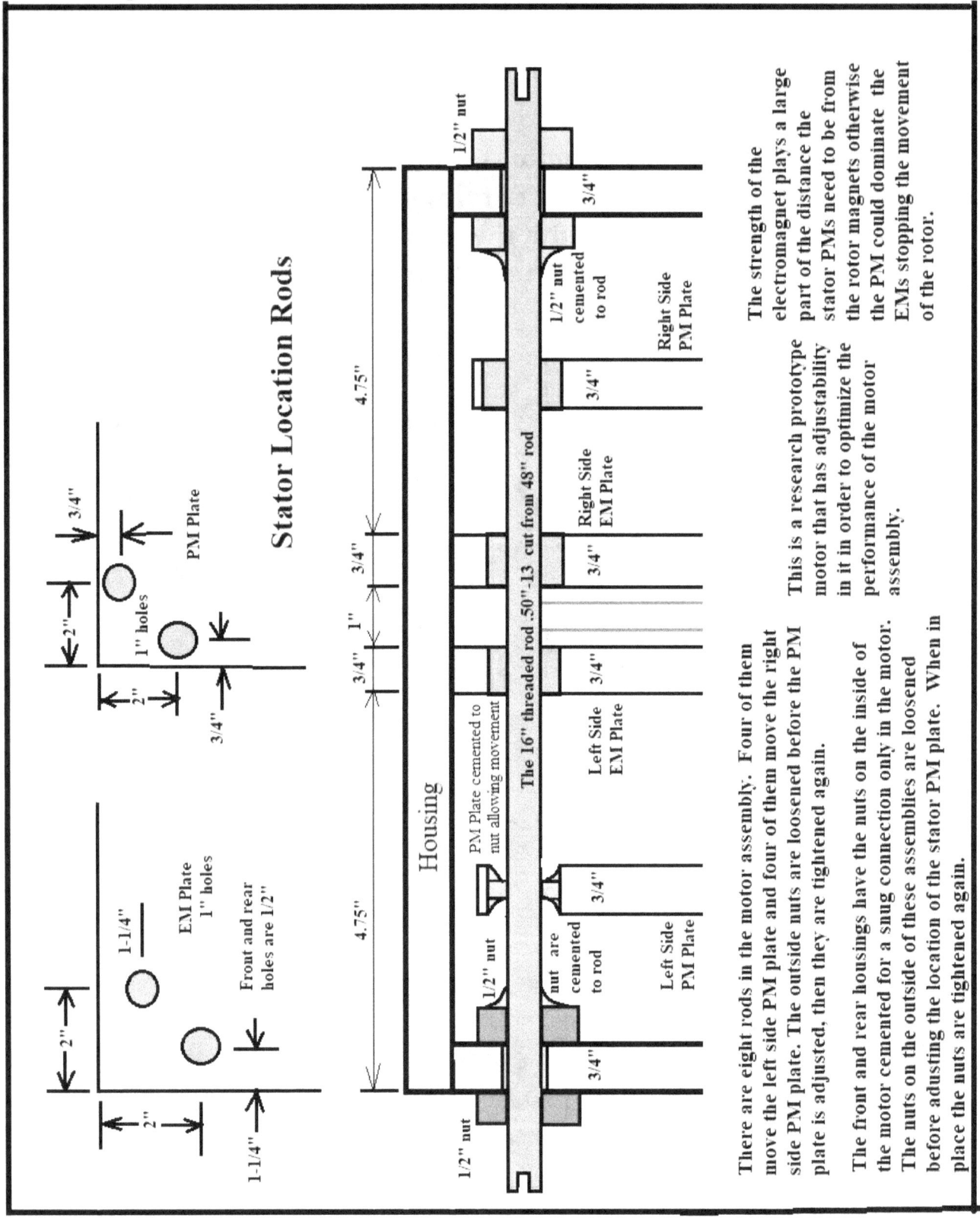

Stator Location Rods

PM Plate

1" holes

3/4"

2"

2"

3/4"

1-1/4"

EM Plate
1" holes

Front and rear
holes are 1/2"

2"

1-1/4"

Housing

4.75"

1/2" nut

PM Plate cemented to
nut allowing movement

nut are
cemented
to rod

3/4"

Left Side
PM Plate

3/4"

Left Side
EM Plate

3/4"

1"

3/4"

The 16" threaded rod .50"-13 cut from 48" rod

Right Side
EM Plate

3/4"

3/4"

4.75"

1/2" nut

1/2" nut
cemented to rod

3/4"

Right Side
PM Plate

1/2" nut

There are eight rods in the motor assembly. Four of them move the left side PM plate and four of them move the right side PM plate. The outside nuts are loosened before the PM plate is adjusted, then they are tightened again.

The front and rear housings have the nuts on the inside of the motor cemented for a snug connection only in the motor. The nuts on the outside of these assemblies are loosened before adjusting the location of the stator PM plate. When in place the nuts are tightened again.

This is a research prototype motor that has adjustability in it in order to optimize the performance of the motor assembly.

The strength of the electromagnet plays a large part of the distance the stator PMs need to be from the rotor magnets otherwise the PM could dominate the EMs stopping the movement of the rotor.

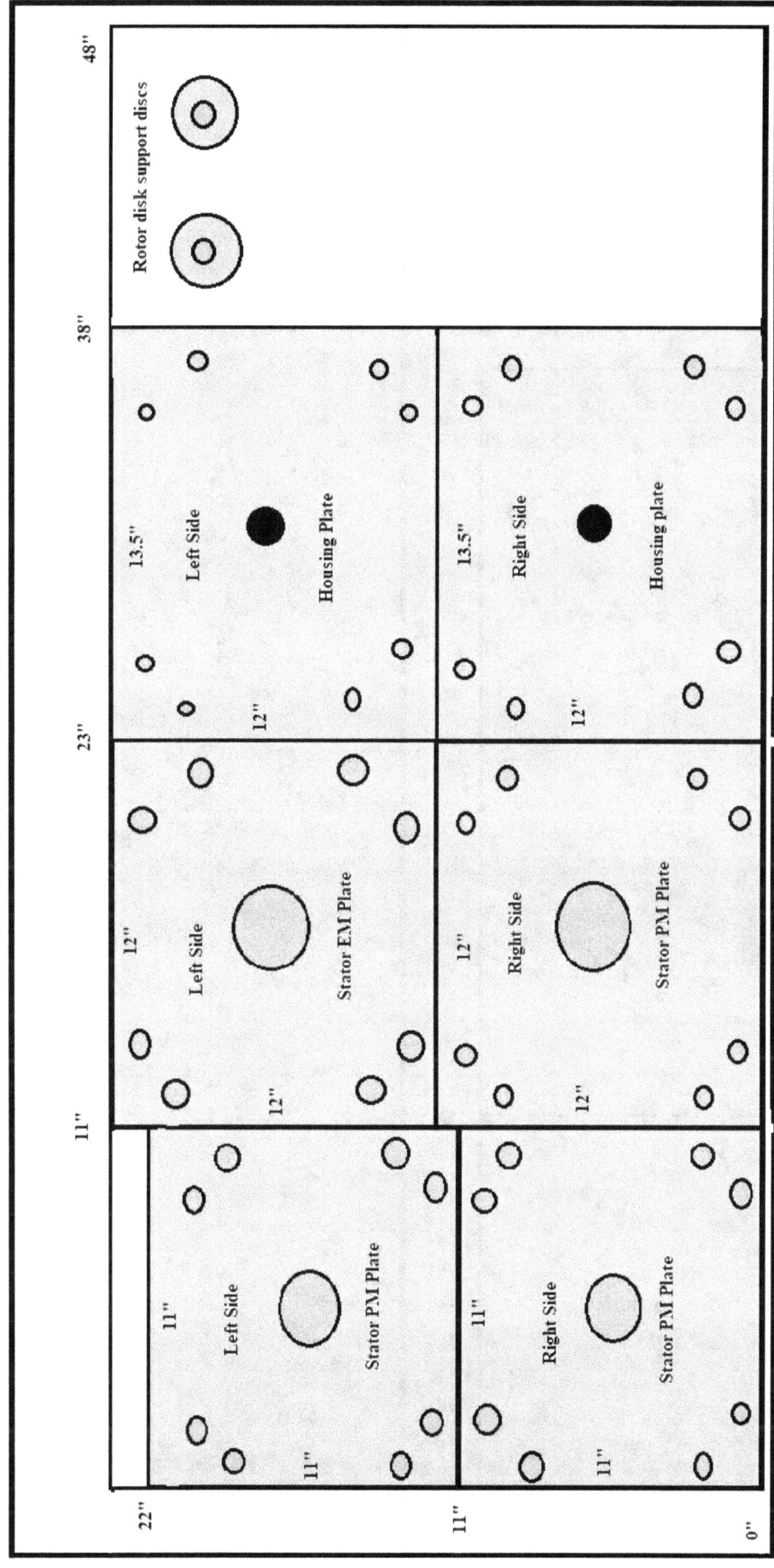

Three Layer Disc Motor Housing Plates: I will not specify mounting parts because they are not as criticle. I do recommend that the near side plate either hinged other easy mechanical access to have the door opened. People need to have this easy access and the ability to have the panel secured to provide safety for the operator. I selected a plastic that is used for motors housings and machine gaurds in several products so that the design will be robust for many years to come.

Top Plate — 13.5", 12", 24"

Bottom Plate — 13.5", 12", 12"

Near side Plate — 13.5", 13.5", 13.5"

Far Side Plate — 13.5", 13.5", 27", 40.5"

Jay Lunke Sept. 22. 2020

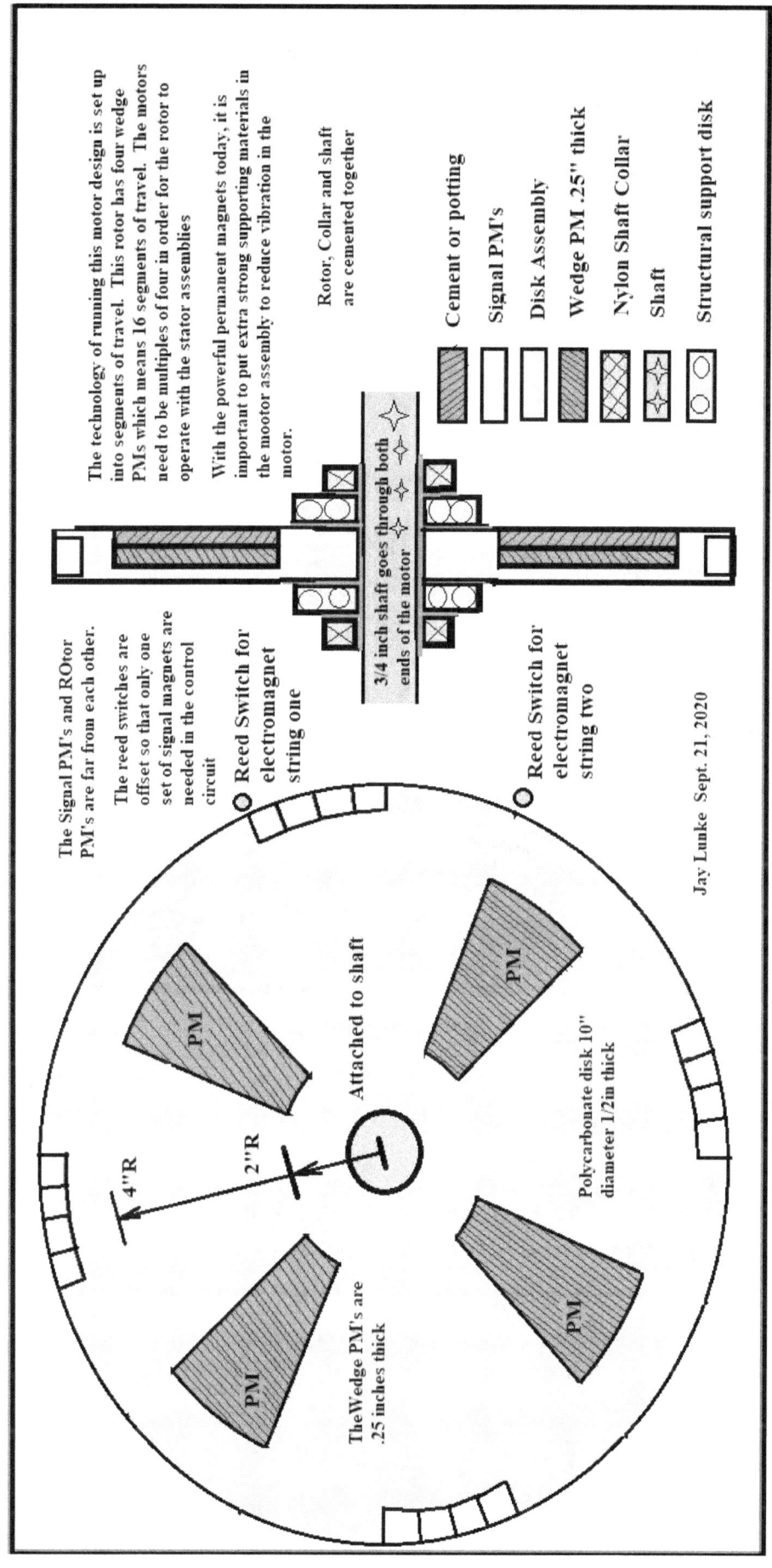

The technology of running this motor design is set up into segments of travel. This rotor has four wedge PMs which means 16 segments of travel. The motors need to be multiples of four in order for the rotor to operate with the stator assemblies

With the powerful permanent magnets today, it is important to put extra strong supporting materials in the mootor assembly to reduce vibration in the motor.

Rotor, Collar and shaft are cemented together

Cement or potting

Signal PM's

Disk Assembly

Wedge PM .25" thick

Nylon Shaft Collar

Shaft

Structural support disk

The Signal PM's and ROtor PM's are far from each other.

The reed switches are offset so that only one set of signal magnets are needed in the control circuit

Reed Switch for electromagnet string one

3/4 inch shaft goes through both ends of the motor

Reed Switch for electromagnet string two

Jay Lunke Sept. 21, 2020

Attached to shaft

4"R

2"R

PM

PM

PM

PM

Polycarbonate disk 10" diameter 1/2in thick

The Wedge PM's are .25 inches thick

25

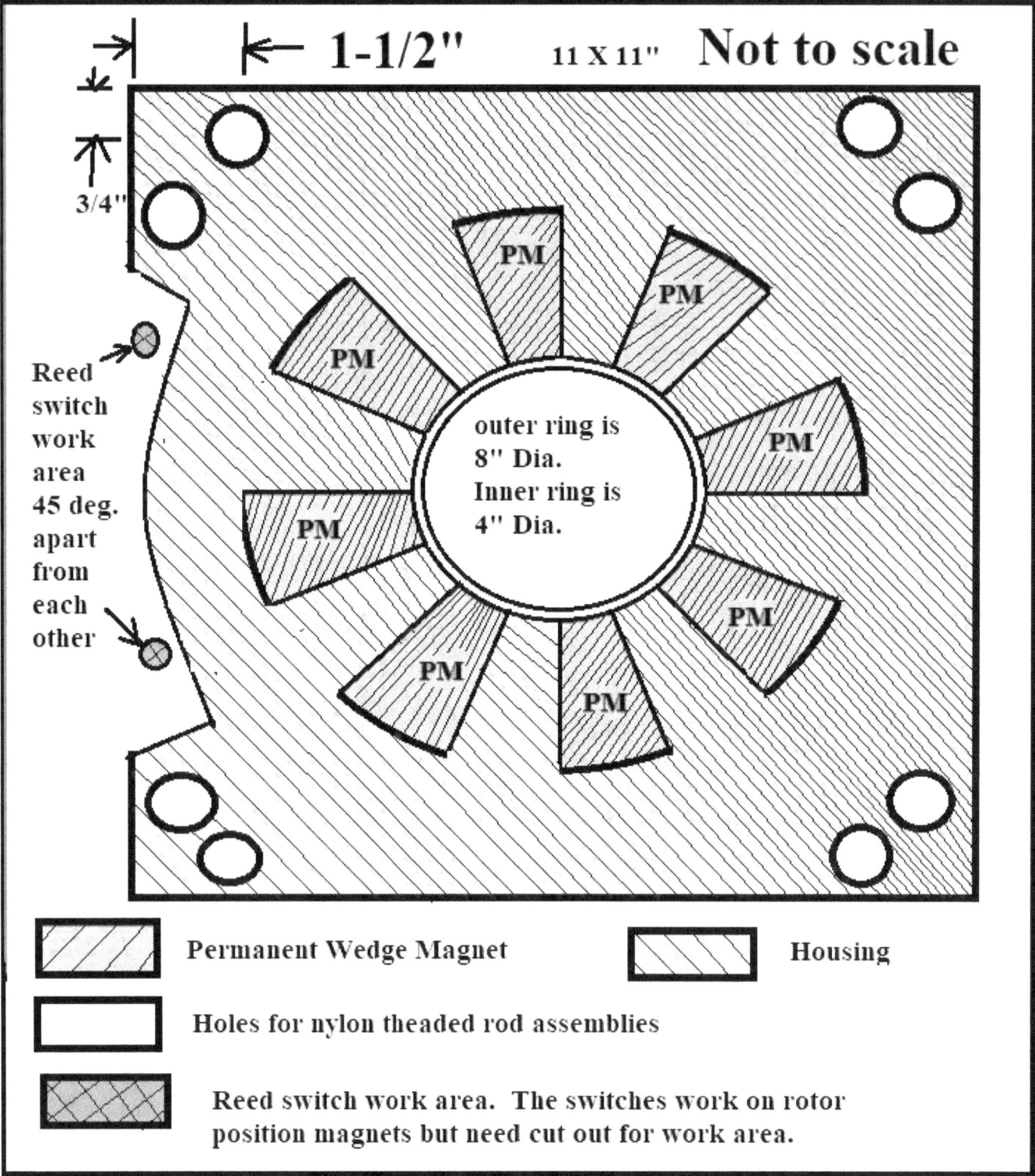

1-1/2"   11 X 11"   **Not to scale**

3/4"

Reed switch work area 45 deg. apart from each other

outer ring is 8" Dia.
Inner ring is 4" Dia.

PM PM PM PM PM PM PM PM

Permanent Wedge Magnet   Housing

Holes for nylon theaded rod assemblies

Reed switch work area. The switches work on rotor position magnets but need cut out for work area.

Stator Permanent Magnet Plate

There are two stator permanent magnet plate assemblies that havr the magnets mounted on the surface of the plate. The magnets face the rotor assembly and the permanent magnets on the plates must line up with each other in order to work. Also the Reed switch work area will be on the other side of the other plate in order to work properly. The Plate and PM's need to be roughed up before cementing them together to make sure they do not seperate during the motor operation. These assemblies are designed to be able to move the PM's into and out of the EM plate assembly.

The motor could operate without these assemblies. These assemblies properly used in the three layer technology of having torque in a ratio of 3:1 of permanent magnets to electromagnets is where greater efficient motor performance happens

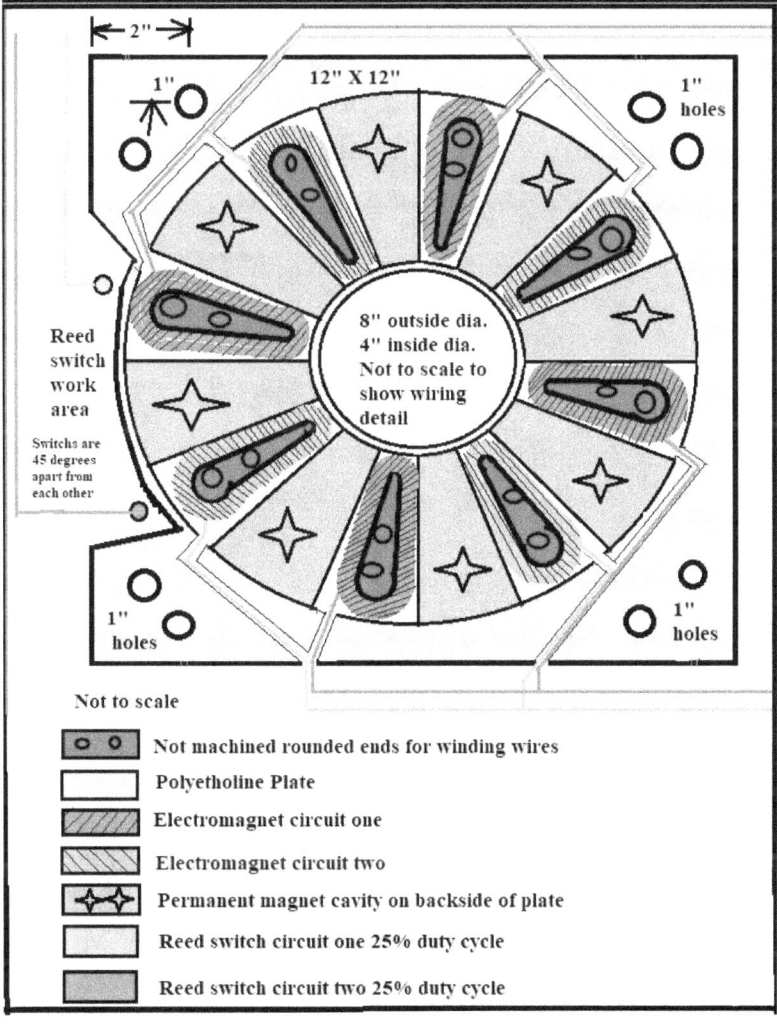

2"

1"

12" X 12"

1" holes

8" outside dia.
4" inside dia.
Not to scale to
show wiring
detail

1" holes

Reed
switch
work
area

Switchs are
45 degrees
apart from
each other

1" holes

1" holes

Not to scale

	Not machined rounded ends for winding wires
	Polyetholine Plate
	Electromagnet circuit one
	Electromagnet circuit two
	Permanent magnet cavity on backside of plate
	Reed switch circuit one 25% duty cycle
	Reed switch circuit two 25% duty cycle

Electromagnet Plate

This is one of two electromagnet plates. The second plate would have the reed switch work area on the right side instead of the left side shown here.

The electromagnet plate is 3/4" thick. The backside has machined areas cut 3/8" deep to accept the PM to slide in from the PM plate assemblies. The EM slots are machined 3/8" deep for the wire assembly to create EM's

Note: Taper PM slots to make sure PMs will slide into electromagnet plate assembly.

Note: The electromagnets of one EM plate assembly MUST be directly across from the electromagnets on the other EM plate assembly.

The following drawing is given to show some specific locations of the permanent magnets and electromagnets on the sub-assemblies. It gives some cautions to watch out for in the build and operation of the mover and power circuit to make sure the prototype functions properly. This motor being built correctly is critical to the proper operation of the motor.

Rotor Moving Clockwise

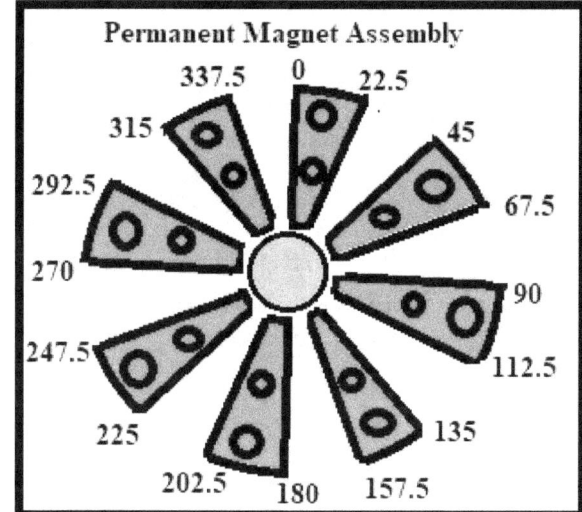

Rotor Moving counter clockwise

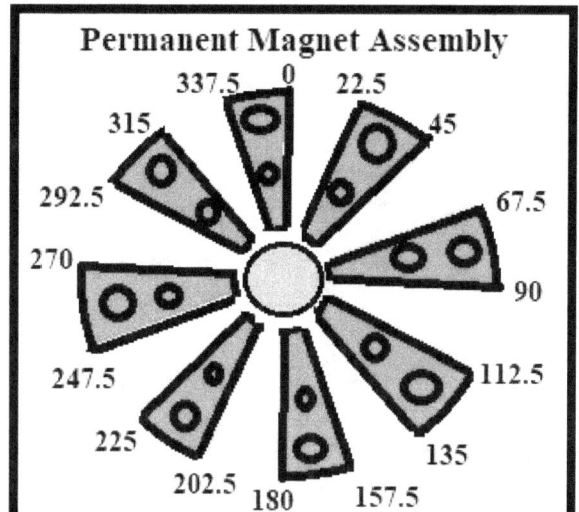

Rotor Moving Clockwise

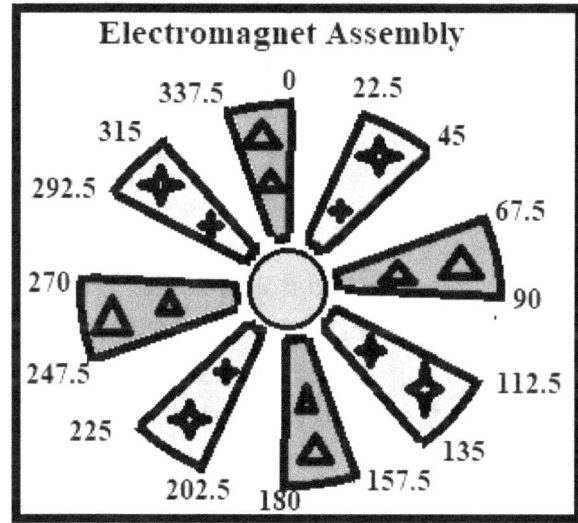

Rotor Moving counter clockwise

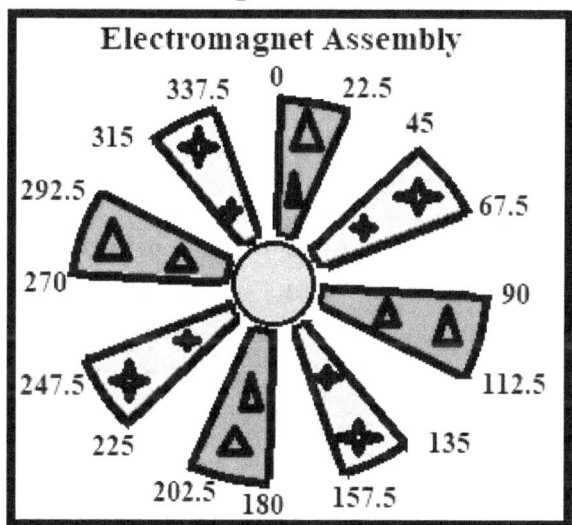

 Permanent Magnets

 Signal Magnet cluster

Electromagnet Circuit One

 Electromagnet Circuit Two

Best Magnet Location for Three Layer Disk Motor

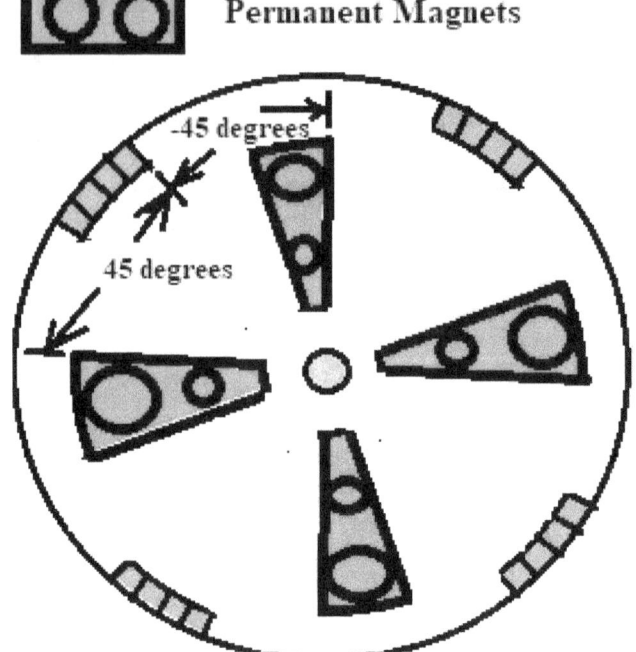

Permanent Magnets

 **Signal Magnet cluster**

Clockwise Side

Reed Switch Nominal Position

Electrical Circuit One options pick 1
Power on at 292.5, off at 315
Power on at 202.5, off at 225
Power on at 112.5, off at 135
Power on at 22.5, off at 45

Electrical Circuit Two options pick 2
Power on at 247.5, off at 270
Power on at 157.5, off at 180
Power on at 67.5, off at 90
Power on at 337.5, off at 0

-45 degrees

45 degrees

Rotor Assembly

It is criticle to to get all of the correct in order for the motor to work. The positions are in different positions on some of my drawings. So check and doublw check these locations out and make sure that assemblies are correct. The clockwise electromagnets need to have the current flowing in the opposite direction of each other because the polarity of the permanent magnet in the rotor is different from one side to the other. The electromagnets dirrectly across from each other need to be turned on and off at the same time to that the forces on the rotor disk are balanced during the operation of the motor. I would hardwire them together to prevent that from happening.

Jay Lunke Sept. 24, 2020

Three Layer Disk Motor Parts List:

Note: The price may have changed since I found them on-line.

1. Wedge Magnet (8" outside Dia., 4" inside Dia., 22.5 degrees, ¼" or optional ½" thick) Note: This is the minimal size that should be used. The magnets need to make a circle in a multiple of 4, these create a circle of 16. Several places sells these. The lowest price I found was about $10 each for lightly blemished magnets.

2. Housing, permanent magnet and electromagnet Plates (Plastic, ¾" thick) Purchase 2 each high density polyethylene sheets 24" X 48" to cut up for the sub-assemblies. I found for $102.76 per sheet on Amazon.com

4. Shaft (¾" dia., 24" long) BearingsDirect has a precision linear shaft for $23.89.

5. Shaft Collar (Plastic, ¾" Bore) 4 each are needed. Zoro has them for $3.38 each.

6. Flange Bearing (Plastic, ¾" Bore) VXB company has thermoplastic ones for $17.79 each 2 each are needed for each end of the shaft.

7. All mounting hardware needs to be plastic (Nylon preferred)

8. ½"-13 threaded Nylon rod, United State Plastic corp. sells these in four foot lengths for $21.11 each. You can cut them into 3 each 16 foot lengths. 8 are needed meaning 3 48" rads are needed.

9. ½"-13 Nylon nuts United State Plastic corp. sells these in packages of 100 for $14.70.

10. 25 gauge copper wire, There are plenty places to purchase motor wire.

31

The above seven drawings are for the prototype build but have similar applications for the 5 to 1 motor. So instead of creating several more drawings for the 5 to 1 motor, use the information in these drawings as a reference for the build of the motor.

ALL PERMANENT MAGNET BUILD, TRAVELING HYBRID FLUX WAVE MOTOR; ADDED AT REV 3

This motor name comes from the fact that the motor design is a re-configuring of the magnets on the stator assembly on the fly in order to create a moving flux wave in it continually in order to keep the rotor assembly move continuously. With the re-configuring, the rotor never crosses a transition point also known as a sticky point at any time during the motor operation. This is unique for any all-permanent magnet motor ever built so far from the research I have done.

I have not built a proto-type motor yet but I am building one now after 50 years of working with designing them. Without a working proto-type, I cannot claim over unity or free energy for my new technologies, that is why I say maybe it has a COP>1. I Have built test stations where I could test modules of the motor but what is needed is a fully operational motor for people to see. I am very excited to be able to be build a full-size proof-of-concept proto-type motor. I have chosen to build a motor design that would have no question of a COP>1 when I am done building it. As a picture is worth a thousand words, a working proto-type is worth thousands of words.

Before I start to explain the motor, I want to explain what this motor is similar to. Picture a man riding a train. The train is on a slope so that the gravity allows the train to move down the tracks. Now there is a limited number of tracks in front of the train. As the train moves down the tracks, the man picks up a section of tracks the train had just moved on and moves it to the front of the train so that it can move farther down the tracks. If there was no bottom of the hill this could go on forever because you never reach the bottom of the hill. What my motor design does is that the stator is constantly being reconfigured as the rotor is moved across it. The magnetism is like the slope on gravity, magnetism will move other magnets to certain positions to make an easier route of the flux to travel in the magnets. My design reconfigures the magnets on the stator assembly on the fly to keep the rotor magnets trying to line up with the stator magnets moving toward the sweet spot where the south pole of the rotor wants to meet the north pole of the stator magnet, but it never gets there. What my new technology does, is to create a forward torque between the rotor and stator through re-configuring the stator assembly through the full rotation of the motor travel. With the gravity driven train, it will reach the bottom of the hill, but with my train, since the magnetic poles keep moving because of the constant reconfiguring of the stator assembly, the rotor will never get to the pole it is trying to reach. The rotation of the rotor never ends because of the constant torque between the stator a rotor assemblies caused by the continuous reconfiguration of the stator assembly. This book will go into a lot of detail of how this works in later chapters. I will only summarize what is happening with the technology here. This chapter gets more into the details of the proto-type motor and hardware I designed and am building. My greatest performing motors I have designed will not be used in this proto-type motor because those designs use electromagnets in the stator assembly in order to operate the re-configuration in them. Those designs do not need moving parts for the switching functions that needs to go on in the stator assembly for the reconfiguration of it. But with electric motors, laboratory testing is required in order to evaluate the efficiencies of them. This would be too costly for me to do. With an all-permanent magnet design without using electricity would be a visual test of its operation. There would be no doubt of its COP>1.

Now every permanent magnet and electro-magnet motor in the world have one or more flux waves in them.

This is because it is the changing flux fields that create movement in them. With that being said most of those designs will have stationary PM and or EM components on both the stator and the rotor assemblies. The electro-magnets will have different currents and directions of currents flow through them in order to generate the changing flux to generate the torque to move the rotor inside of the stator assembly. These methods I will call two-layer systems. What I have been working with for the last few years are three-layer motor designs. What is different about three-layer designs is that the configuration of the stator changes while the rotor is rotating inside of the stator. This can be done either mechanically or electrically. If I, do it electrically, usually I will use an air core for the electro-magnets or coils I am using in the motor. The reason I have been using an air core is because I want the component to disappear from the functional sight of the permanent magnets in the motor at certain times of the operation of the motor. It is because of the electromagnets appearing and disappearing functionally in the motor, I am able to change the functional make-up of the stator without moving hardware during the operation of the motor assembly. When the electro-magnet comes to life, it becomes a part of a hybrid magnet with the two adjacent permanent magnets next to it. This new hybrid magnet now has a different function with the permanent magnets on the rotor assembly. It is the cycling between the magnet to hybrid magnet arrangements that creates a condition where the rotor permanent magnets will always have a forward torque from the stator assembly to the rotor assembly. In this book, I use this method of design in many of my motor's designs. In my latest designs I have used physical moving parts in order to reduce the number of electro-magnets in the motor. I feel that the fewer electro-magnets I use then the less power will be needed to operate the motor. Now mechanical movement creates more losses that need to be taken into account as to the final performance of the motor. Once the technology is proven with mechanical movement, then the electrical motor designs will be explored with more enthusiasm and expectation of positive results in those motor designs.

Now what I have achieved by using the three-layer motor designs is to reconfigure the stator of the motor on the fly. When working with an all-permanent magnet design and trying to get continual movement between them there is that hump of "for every attraction there is a repulsion" between the two magnets that needs to be resolved. This is why in a lot of conventional electric motors there is a permanent magnet working with an electromagnet to control those negative torque points in the motor performance.

With the three-layer technology, let's say we are using two permanent magnets again, one in the stator and one in the rotor. Now let's assume that attraction is being used to create forward torque in my design. Now I would use the attraction of the two magnets to rotate the rotor half ways through the rotational travel from only the force existing between these two magnets, the magnets would now resist the movement of the rotor because of the repulsion between those two magnets. A conventional motor would have replaced one of the permanent magnets and used an electro-magnet in its place in order to provide a current in the opposite direction in order to move the rotor the rest of the rotor's rotation. What you have to do is have electrical energy in the electro-magnet through the full rotation of the rotor now. So, this means you have torque from 50% permanent magnets and 50% electromagnets. What the three-layer technology does is to reconfigure the stator into a functional magnet that creates forward torque from the stator assembly hybrid functional magnet and the rotor permanent magnet. This is the place where the normal repulsion occurs in the two-layer design. So, 50% of the rotation is done in the two-layer format and the other 50% of rotation is done with the three-layer format. So, you end up having torque from the permanent magnets and one hybrid functional magnet for the full rotation of the motor. The functional operation is through 50% of the rotation. When using the electromagnets in the design, you are operating at less power than other motors.

When using electrical energy, enough flux needs to be generated to create the hybrid magnet.

Now as the rotor moves inside the stator assembly using this alternating of hybrid to conventional configurations, then the device will have flux waves that are created and will move the rotor around in a circle during the operation of the motor design. Now there will be one flux wave for every four segments of rotor travel. If the motor has 32 segments to create one full rotation of the rotor, then there will be eight traveling waves in the motor.

Now for designs that use mechanical movement to make up the hybrid configurations, mechanical energy is needed in order for this operation to occur. The stator configurations need to be in sync with the rotor's rotation at the same time. Now there are endless ways to mechanically configure the hybrid and disassembly of it. What I have decided to use in my motor design is a switching wheel with eight segments of configuration on it.

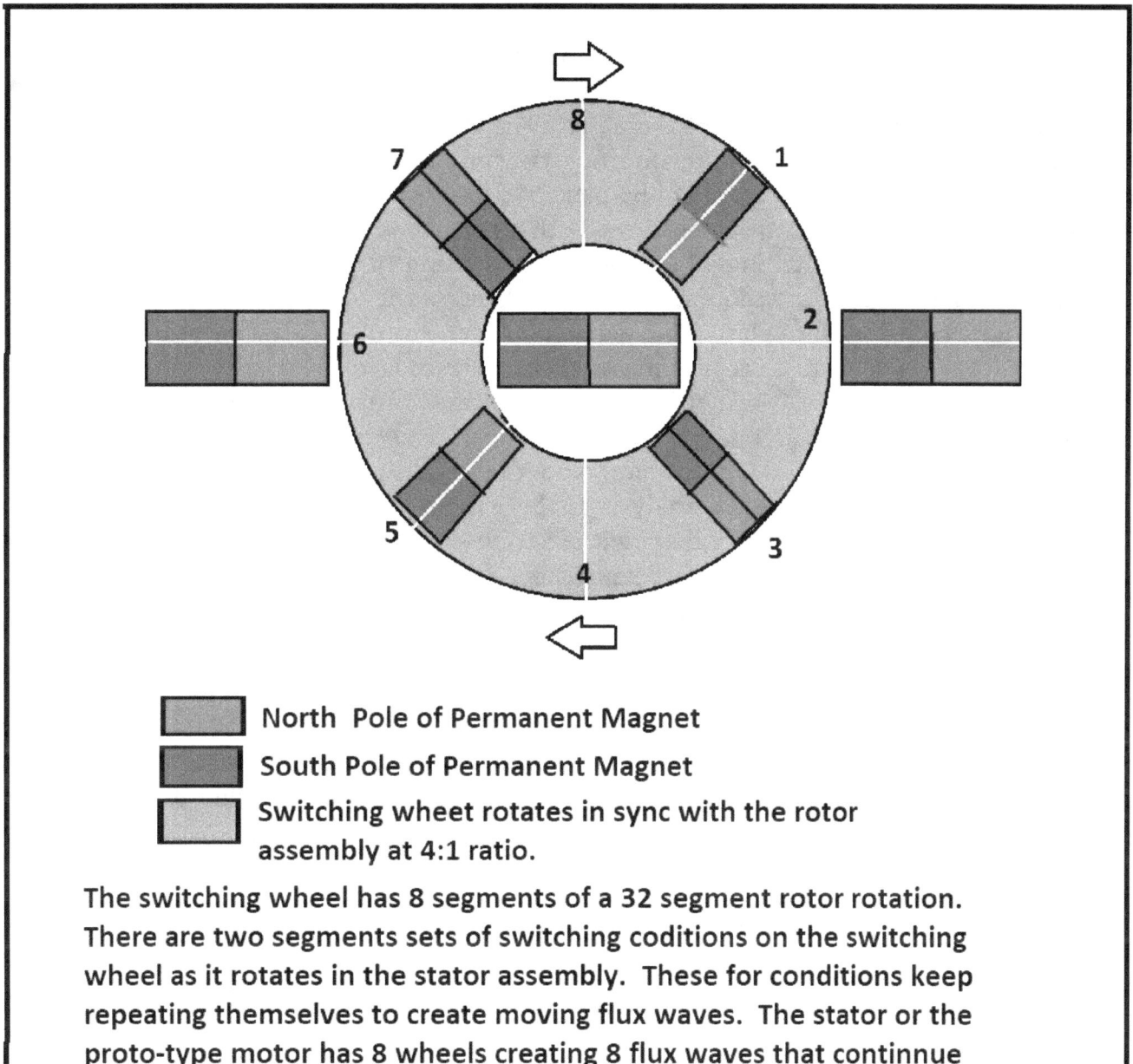

North Pole of Permanent Magnet

South Pole of Permanent Magnet

Switching wheet rotates in sync with the rotor assembly at 4:1 ratio.

The switching wheel has 8 segments of a 32 segment rotor rotation. There are two segments sets of switching coditions on the switching wheel as it rotates in the stator assembly. These for conditions keep repeating themselves to create moving flux waves. The stator or the proto-type motor has 8 wheels creating 8 flux waves that continnue to rotate around in a large circle in order to create a motor that has a rotor that rides these moving waves. THe moving waves create torque 360 degrees in the rotor for each of the 8 permanent magnets. As the switching wheel rotates, it's permanent magnets create attraction and repultion with the stationary permanent magnets of the stator. These forces nulify each other and reduce the friction on the switching rotor bearing assembly.

Each switching wheel works with a pair of switching positions along the stator assembly. I call one segment of travel as the length of one magnet on the stator assembly. The switching rotor is called out in segments of travel as well. There are four segment configurations on the switching wheel that are repeated again to create the eight segments on the switching wheel. Now what I have decided to use for my switching wheel is magnets for the switches The switches will not only create the hybrid magnet on the stator assembly when needed to produce rotor travel but it will also place a magnet of the opposite polarity on the other end of the hybrid in order to strengthen the interaction of the hybrid with the rotor magnet. The way it is strengthened is that it helps to prevent the stator magnets into becoming one large ring magnet because of the stator permanent magnets close proximity with each other. By pairing two switching magnets going into switch positions having opposite magnetic poles to each other is a big advantage in reducing unwanted torque and friction on the switching wheel. As one switching magnet is being placed into the switching position it will have attraction to move into place. At the same time the other switching magnet will resist being put into place. The net result for the most part will be a cancelation on the switching wheel. Then when the first magnet is being removed from its switching position, it will resist that movement. While this is happening the other switching wheel magnet is being pushed away from the switching position. The forces between the two switching magnets again for the most part cancel each other out. Now the reason I said mostly cancel each other out is because if you perform a vector analysis of these movements, there will be some overall resistance to performing this function along with friction in the bearings, wind resistance of the spinning switching wheel and other misc. performance reductions that become a factor. Now in the final motor assembly, the rotor magnet will also contribute some additional reduction to the performance of the switching wheel.

A series of the switching wheels can be added together to create a track for linear movement. What I do is to bend my switching assembly for the rotor to move in an arc. I put 8 of these assemblies together so that the eighth section connects to the first switching network again. I end up with a high torque rotational motor design. This works good for the traveling flux wave to have a continuous path to travel.

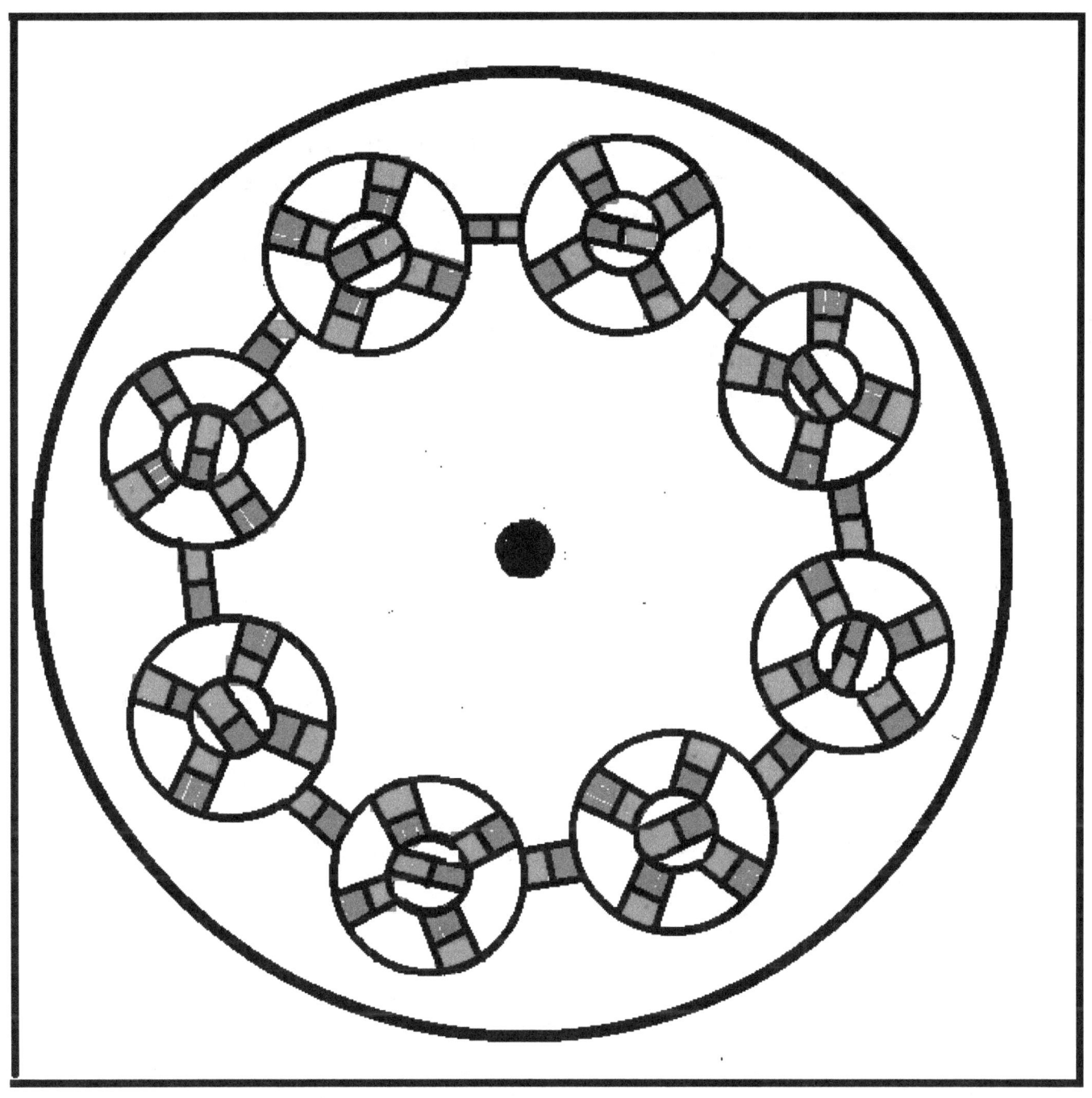

Stator Assembly

The Stator operates like a wave pool pushing 8 flux waves around in one big circle as the 8 switching wheels rotate to generate these waves.

The rotor permanent magnets have forward toque on them through all of the rotation of the rotor as a result of this action like a person surfing on an ocean wave or a dog chasing its tail.

The electromagnets can either operate at a 50% duty cycle for forward torque when needed for the motor or by switching the dirrection of the current through the coil at the correct time to generate forward torque throughout the rotation of the rotor assembly.

The trade-off; The more the stationary magnets are lined through the switching magnets, the better the ballance in the switching wheel will be. THe farther away the switching magnets are from the stationary magnets in the off condition, the better the performance of the motor.

WHen the functional magnet becomes larger, the less torque is between the rotor and stator magnets will be.

Eight flux waves are created and will flow around the large circle in sync with the rotor movement because of the gears are driving the switching wheels

Now as the rotor moves along the tract of the stator there is always torque between the stator and the rotor assemblies supporting the rotation of the rotor. Now the rotor is connected through a gearing system to each of the switching wheels. Since the stator has 32 segments in one full rotation of the rotor and the switching wheel has eight segments of function on it, the switching wheels will rotate four times around for each rotation of the rotor. So, the four to one gearing ratio between the rotor and switching wheels will create a one-to-one segment movement between the stator and rotor assemblies. The segment positions between the rotor and switching wheels are critical. So that is one reason there can be no slippage of the main shaft to the rotor disk, main shaft gear to main shaft, switching wheel to switching shaft and the switch wheel gear to switching wheel shaft. Since there will be torque on these components, then extra design thought needs to go into them. Now the main 6" gear has 192 teeth in it. The next 3" gear has 96 teeth in it. The 1 ½" gears have 48 teeth in them. This means that gong from the 6" gear to the 1 ½" gear will have a ratio of four to one. The more teeth the better the synchronization will be because there will be less gear slop in the motor. I used 17 gears instead of 9 gears in order to save on cost for this proto-type build. It just so happens that the extra layer of gears causes the rotation of the rotor and switching assemblies to be in the same direction. Functionally, this should not make any difference.

Since a picture tells a thousand words, I am adding pictures of the proto-type here to show how the rotor and stator can work together for this new technology.

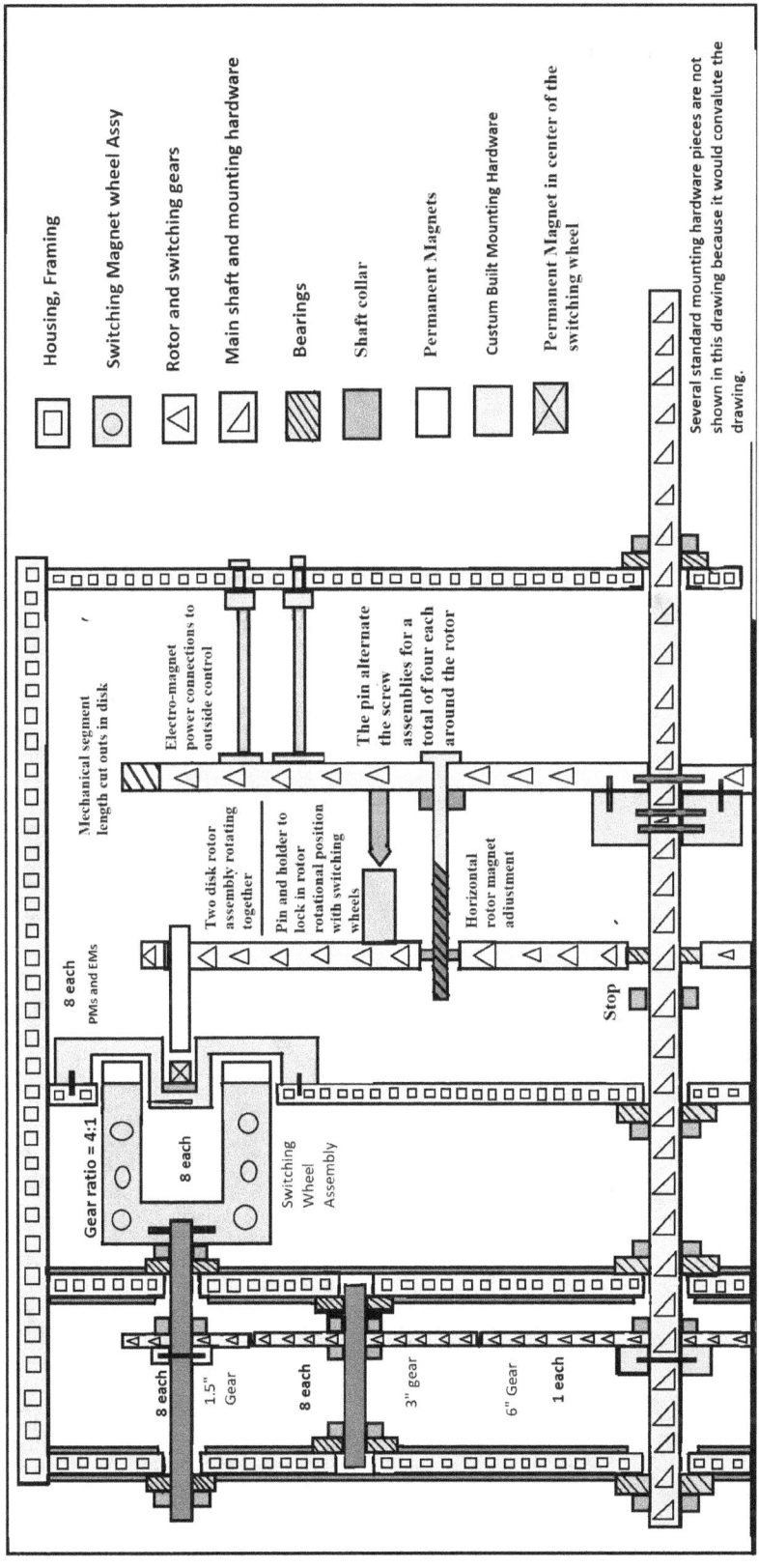

This drawing shows connection for an external power supply to drive the optional electro-magnets if needed to power the motor. (Not to scale)

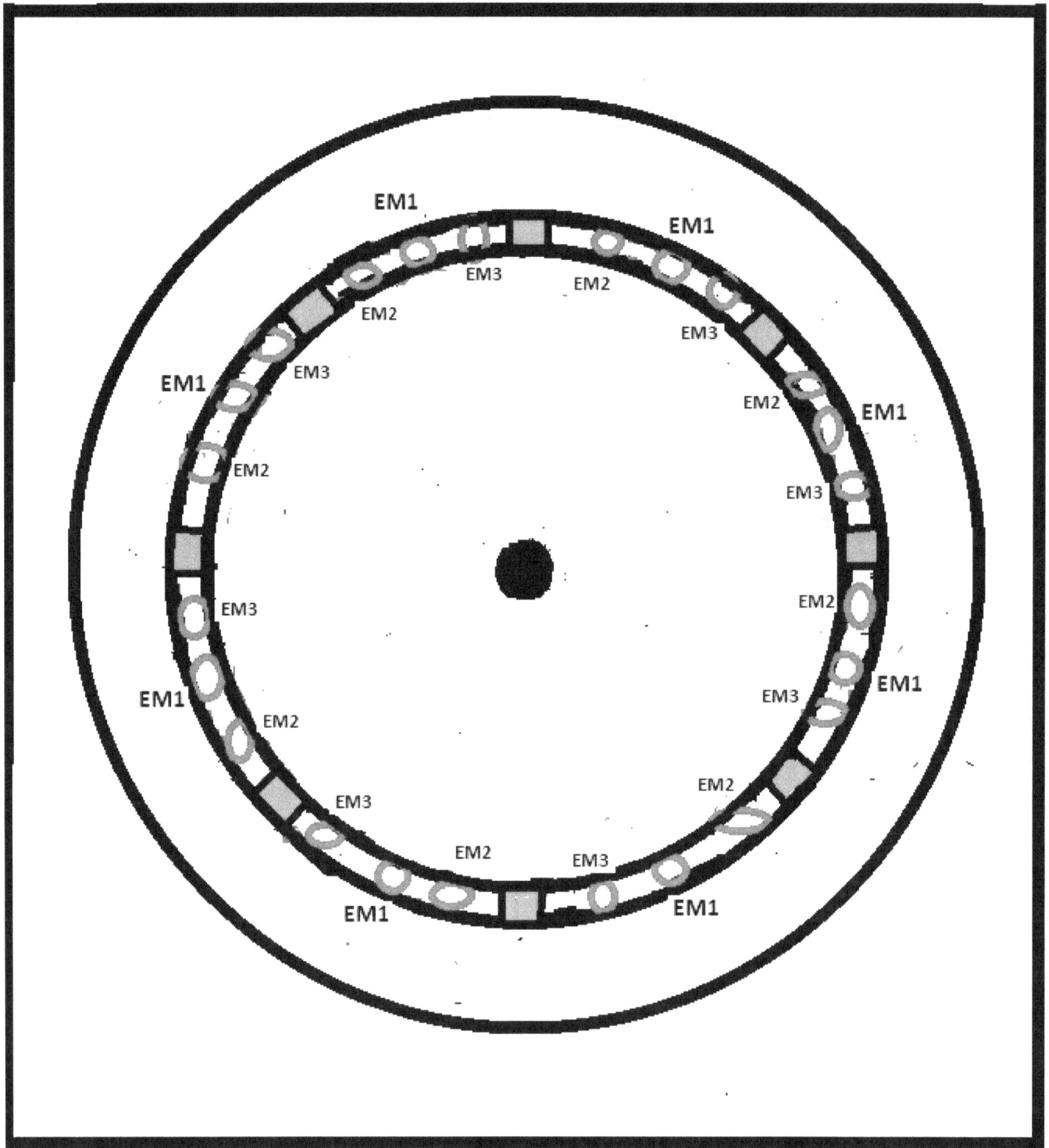

Rotor Assembly

Electromagnet Value to motor

EM1 has the most benifit to the motor

EM2 and EM3 have less additional benifit and will not be built into the functional proto-type motor.

When using dual power supplies, the current can be switched through the electro-magnet in order to produce forward torque in the motor through the full rotation of the rotor assembly.

▨ Permanent Magnets

◯ Optional Electromagnets

Note: The Permanenet magnets have forward torque on them all the time so the duty cycle is 100%.

When using a single power supply

The elctromagnets are 50% duty cycle and are turned on only optionally when the flux from the EM produces forward torque on the motor assembly. This is not the case when using the EMs for breaking the motor.

modified 12-7-20 Jay Lunke

Since this proto-type will be able to test a few different concepts with it. This drawing shows the rotor makeup of permanent magnets and electromagnets.

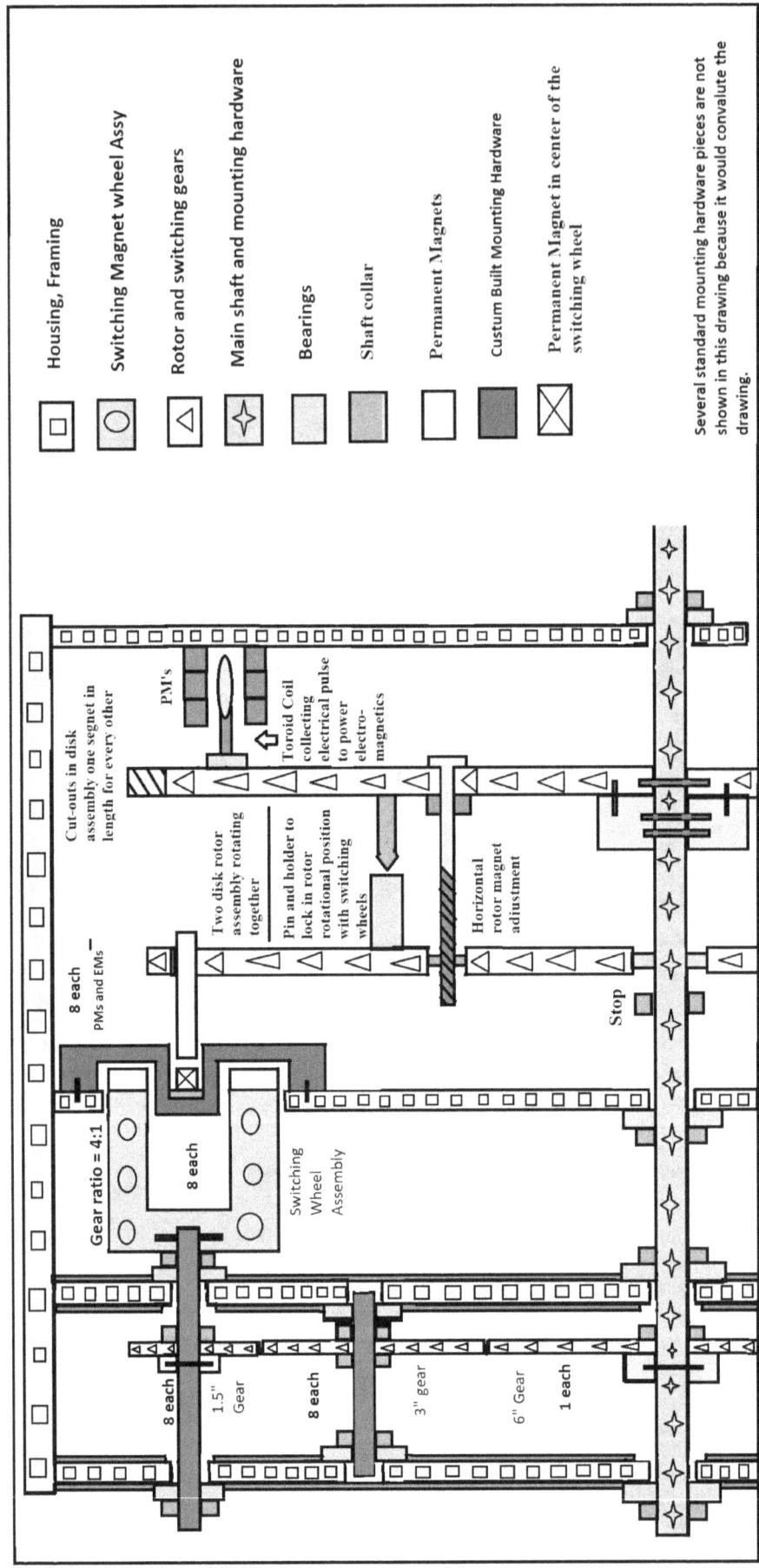

This drawing above shows both the motor and the on-board generator to drive the electromagnets. (not to scale)

Now the design of the on-board generator could be evaluated with the dual pendulum approach in order to reduce the resistance the generator coils would have in the motor. The toroid or partial toroid coils moving through a magnetic field to pulse the rotor electromagnets at the correct timing is the first this I will be testing. I have a two coil using air coils that function like a toroid when the coil pair moves through a magnetic field. Of course, the final design needs to alternate a positive current in one segment with a negative current in the next segment and repeat that sequence through the remaining rotor rotation.

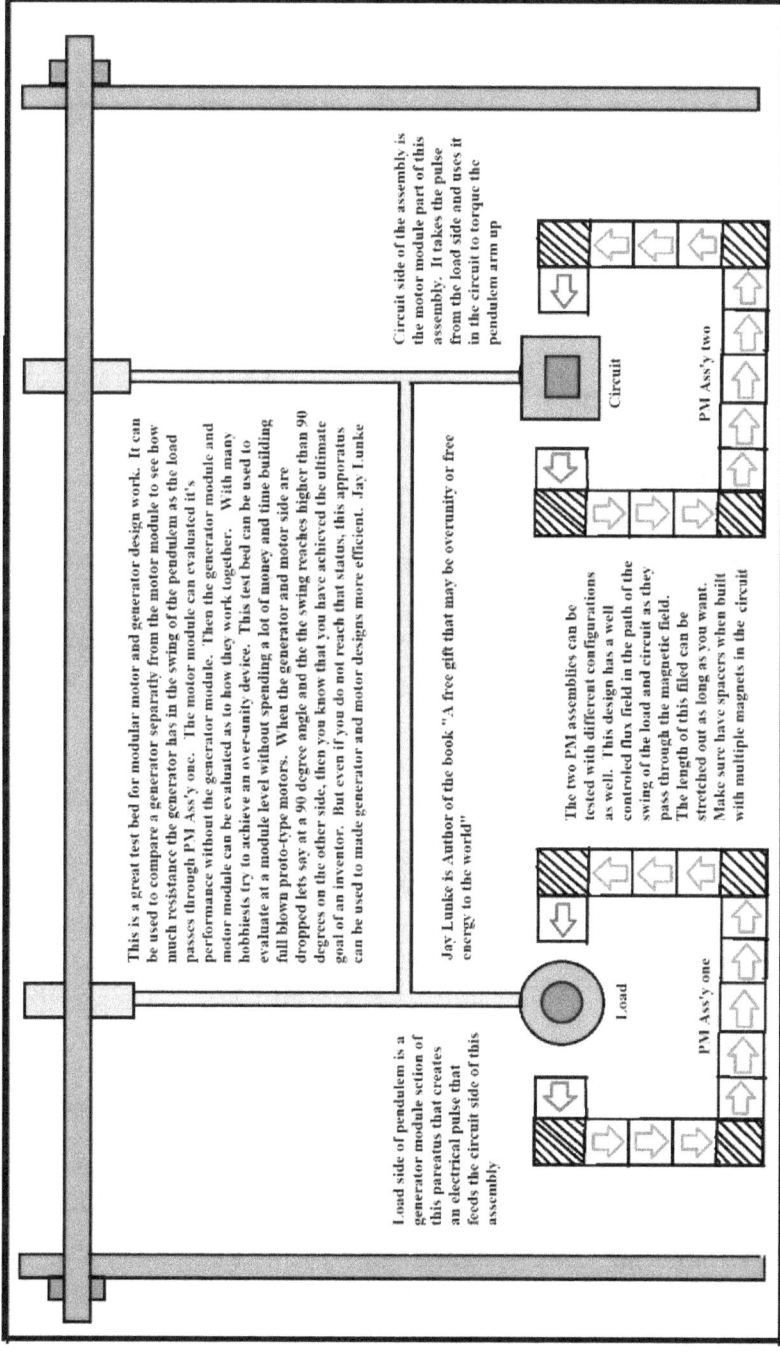

This is a great test bed for modular motor and generator design work. It can be used to compare a generator separately from the motor module to see how much resistance the generator has in the swing of the pendulum as the load passes through PM Ass'y one. The motor module can evaluated it's performance without the generator module. Then the generator module and motor module can be evaluated as to how they work together. With many hobbiests try to achieve an over-unity device. This test bed can be used to evaluate at a module level without spending a lot of money and time building full blown proto-type motors. When the generator and motor side are dropped lets say at a 90 degree angle and the the swing reaches higher than 90 degrees on the other side, then you know that you have achieved the ultimate goal of an inventor. But even if you do not reach that status, this apparatus can be used to made generator and motor designs more efficient. Jay Lunke

Jay Lunke is Author of the book "A free gift that may be overunity or free energy to the world"

The two PM assemblies can be tested with different configurations as well. This design has a well controled flux field in the path of the swing of the load and circuit as they pass through the magnetic field. The length of this filed can be stretched out as long as you want. Make sure have spacers when built with multiple magnets in the circuit

Load side of pendulem is a generator module section of this pareantus that creates an electrical pulse that feeds the circuit side of this assembly

Circuit side of the assembly is the motor module part of this assembly. It takes the pulse from the load side and uses it in the circuit to torque the pendulem arm up

Circuit

Load

PM Ass'y one

PM Ass'y two

Now the dual pendulum will be used to optimize the best on-board generator for me to power the electromagnets of this motor. The final proto-type will have both on-board generator and connections for external power supplies to operate electromagnets. Yes, the first proto-type will be kind of a test platform. Even the rotor is built with two layers to adjust the height of the rotor magnets. All of these things will help to optimize the design for the next level of proto-type that would require a machine shop to build it.

Now even if the all-permanent magnet version operates the motor, I want to add the electromagnets to increase the power of the motor. With the way and amount of power is injected into the electromagnet, the motor can have a wide range from a stopped position to a fast rotation. The rotation cannot be two fasts because the switching wheels would fall apart at higher speeds.

The following few drawings show some different options for developing the motor. There are simpler designs in the book that require electromagnets in the design and they will be less expensive to build. So please be patient with these designs for now.

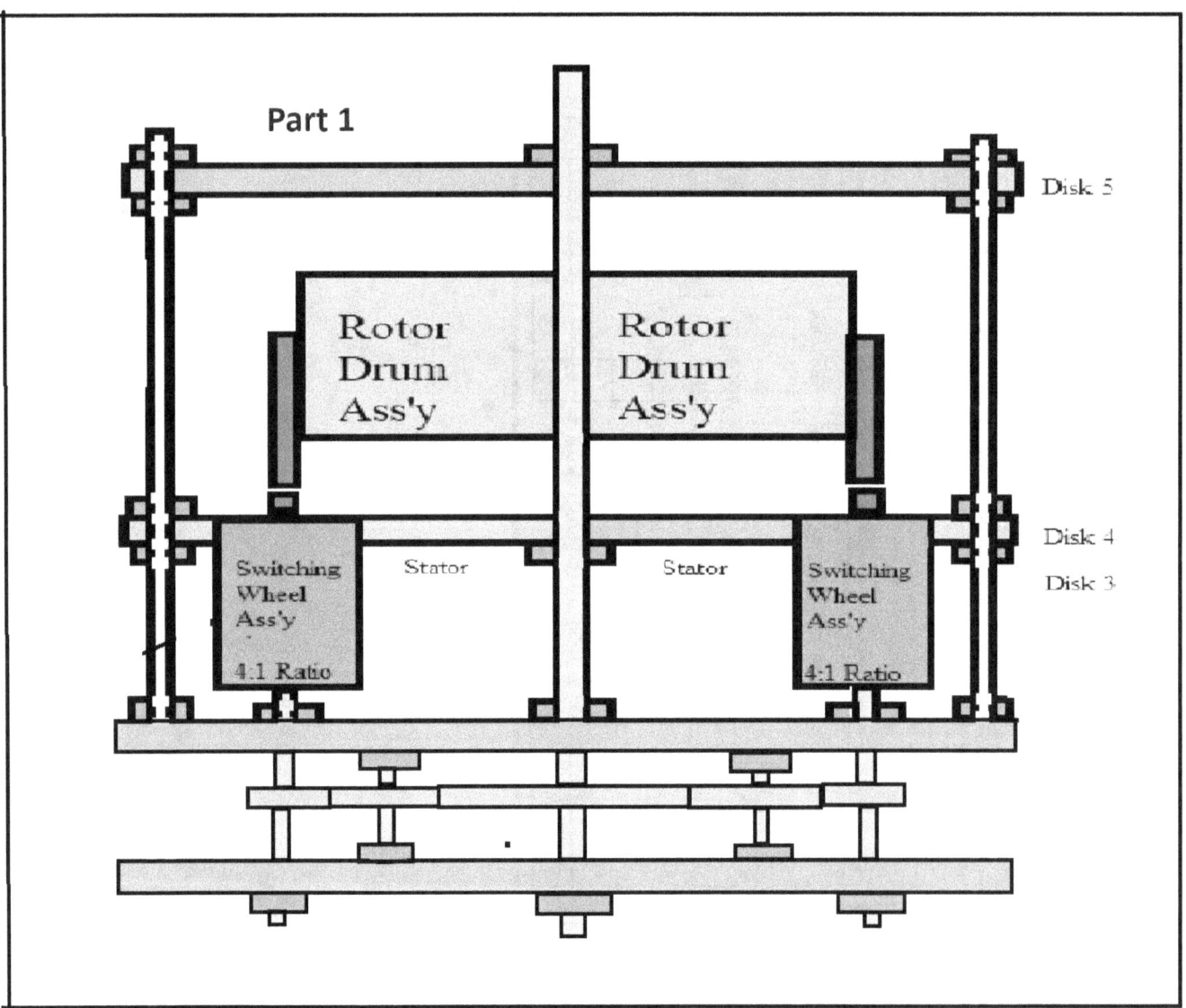

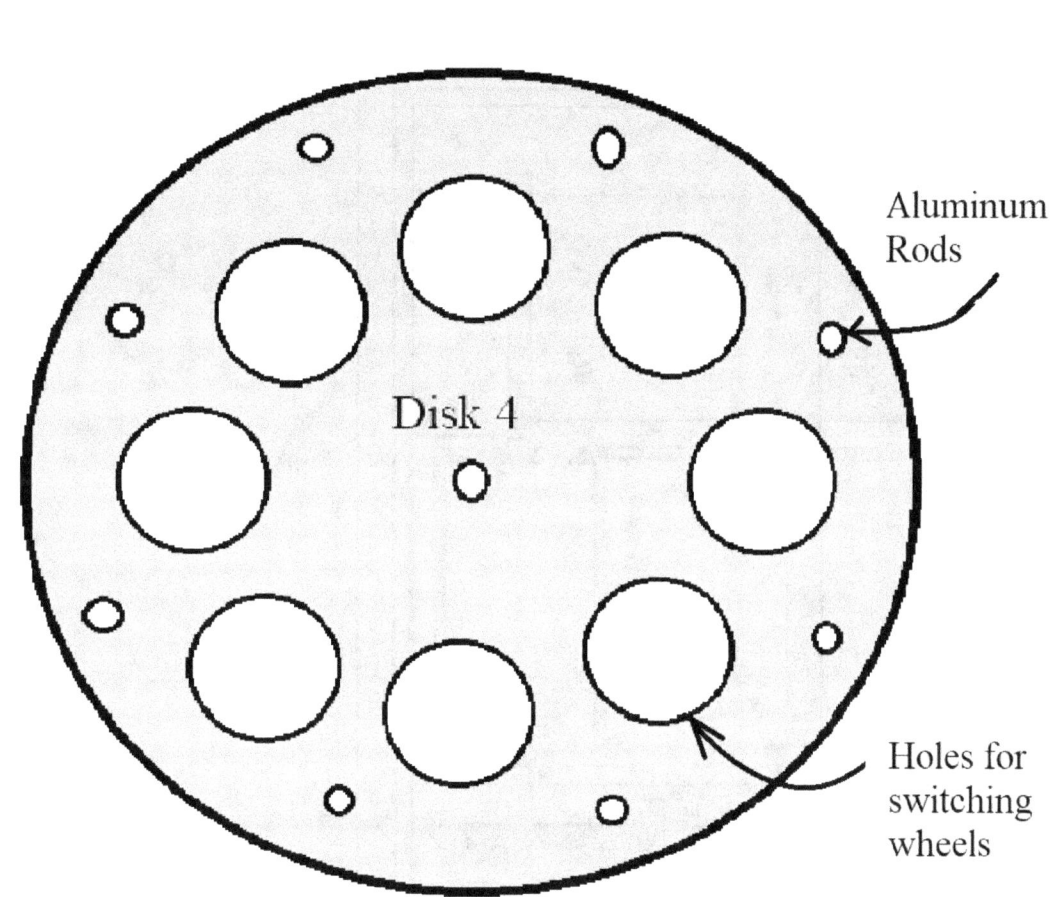

The Drawing mainly shows that there are eight switching wheels that are
spinning at a ratio of 4 turns to one of that of the rotor assembly in order
for the eight segments of functin on each switching wheel mate up with the
32 sements of function on the rotor assembly. This is what will allow the
rotor to conntinnue to to keep in sync. with the switching assemblies to
continnue to create and rotate eight flux waves in the stator assembly for as
long as the motor is rotating. The Rotor PM's surf on these waves.

This drawing shows more information of how the hardware and framing are built in the proto-
type.(not to scale)

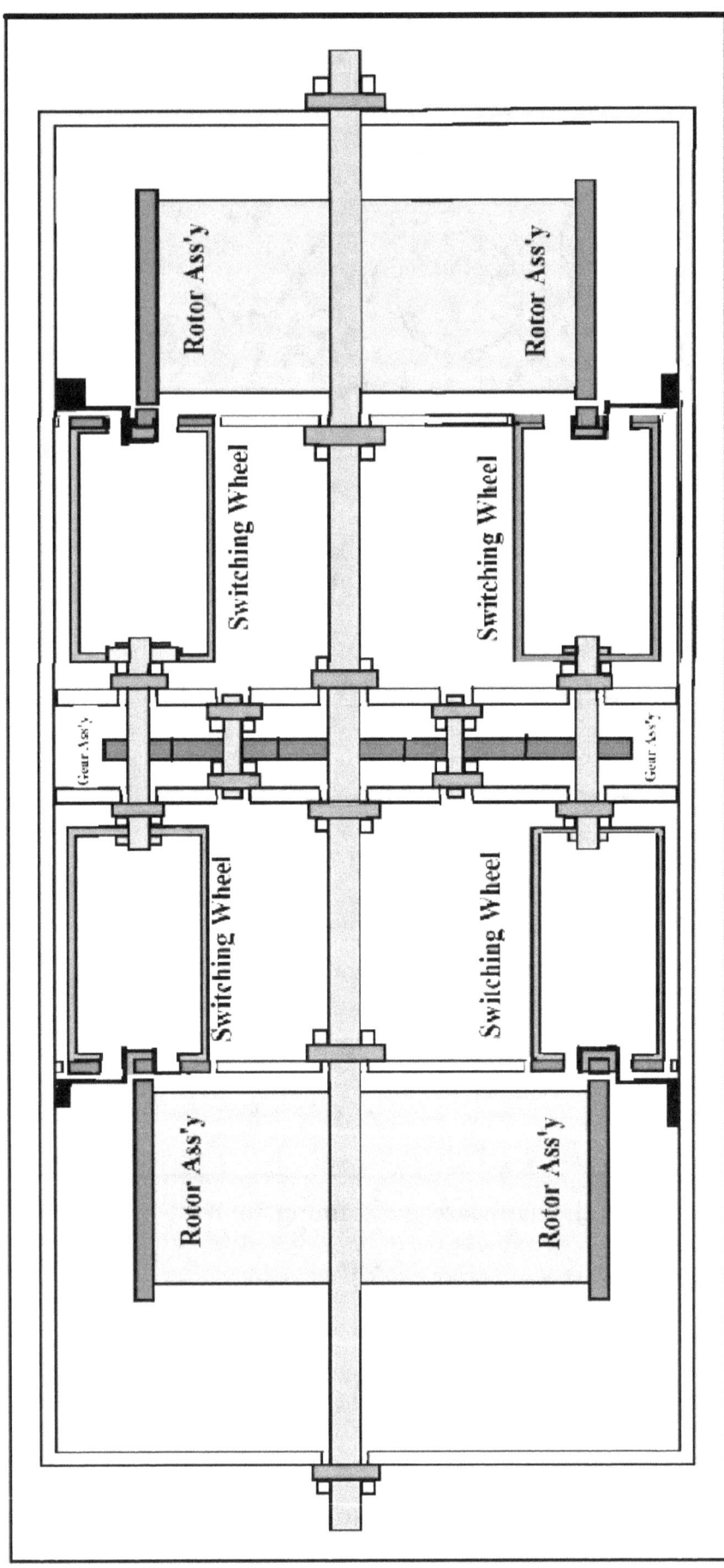

This drawing above only has the basics to show that a common gear box can be used to drive two motor sections. The polarity on the stator would need to be switched on one of the motors in order to have the torque on the rotor in the same direction. (Note to scale)

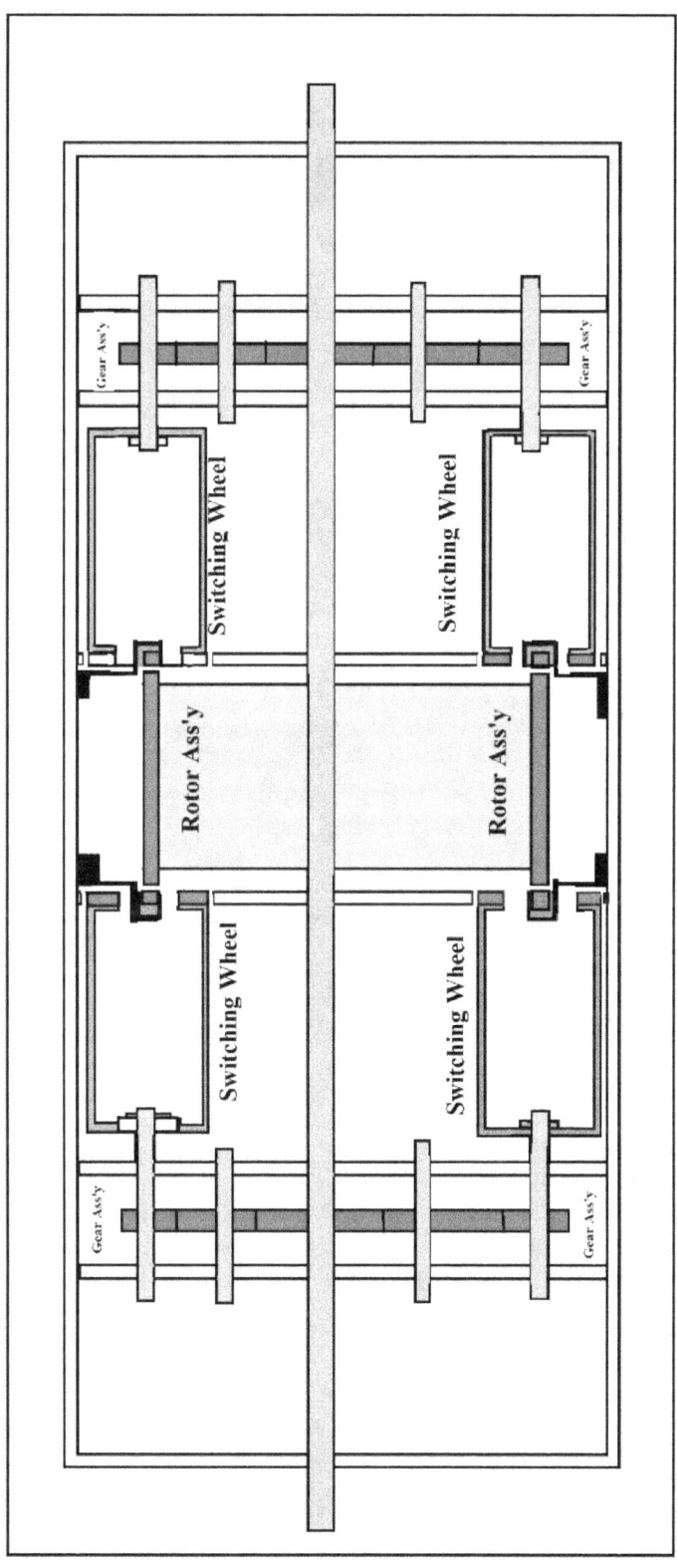

This is another option of using a common rotor assembly. This saves on permanent magnets needed if two motors were built to increase power output of the motor. (Not to scale)

Now with all of the options shown above, they all have a traveling wave in the stator assembly that the rotor travels along the sweet point of that flux wave generating the movement of the motor. In the following drawings do not let the backwards arrows deceive you as to resistance to the motors movement because it builds the strength of the forward torque of the flux wave.

The following drawing shows the hardware configuration flux activity during the four segments of the stator using the "Three Layer Electro-Mechanical Movement Technology. The drawing after that shows the path the rotor magnet needs to be at for the operation of the motor to work. Both the proper mechanical positioning and electrical execution need to be correct for this motor to work properly. The other chapters will give you a better description of how this operates.

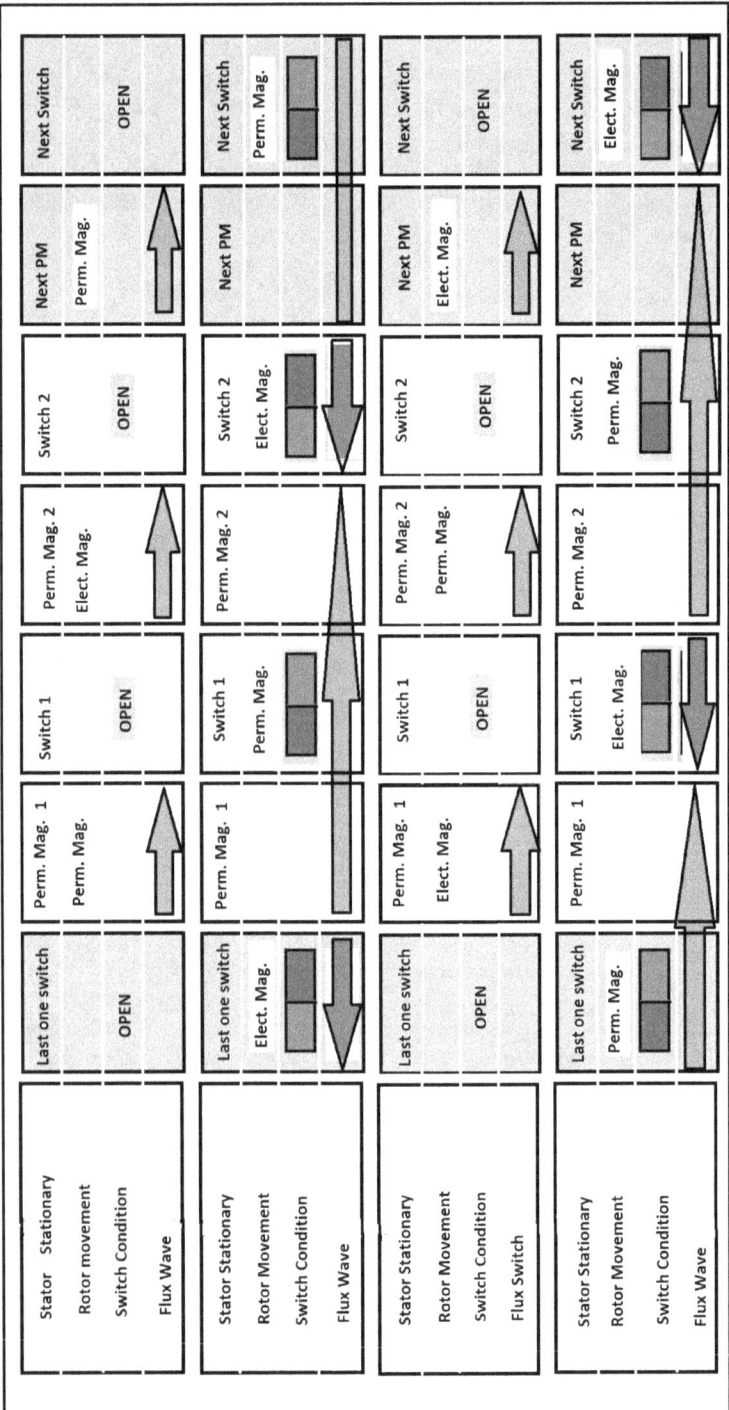

The rotor PM's ride on the angular green bars of this graph. It is the angled shaded areas in the drawing. Between the green or shaded bars, the electromagnets take advantage of cycling the power of the electromagnet to keep forward torque on the rotor throughout the full travel of the rotor assembly.

The key to understanding the movement of these motor types is to know that there is one rotor magnet per four segments of stator assembly. So as the rotor magnet moves through the stator assembly, it basically is only one segment at a time. So, I move the rotor through the stator assembly so that the magnet stays in the shaded spot through the full rotation of the motor operation. After the rotor moves through the four segments of travel, then the process repeats itself for the next four segments of travel.

The solid arrows show the forward torque while the white arrows show the reverse torque. Now I make sure never to have the rotor magnet in the segment of the stator when that segment is configured with reverse torque.

These two drawings are unique to the "Three Layer Electro-Mechanical Movement Technology".

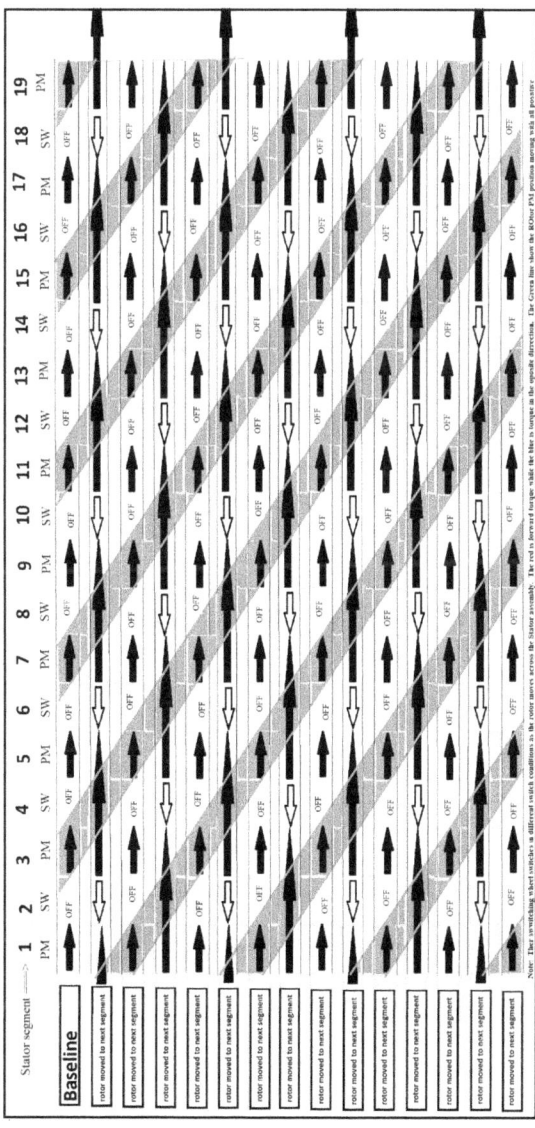

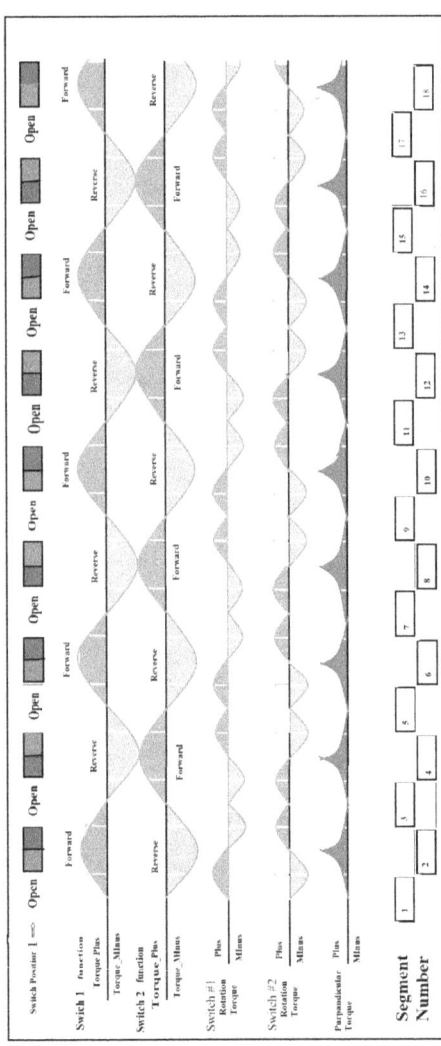

The next graph is more technical because it summarizes the torques and resistance to the movements of the switching assemblies as they apply to the final overall performance of the motor assembly. These two drawings above and the following drawing are for motor designs using permanent magnets in the switching assemblies. These permanent magnets cannot be turned on or off. This results with more complicated designs. Now the switching assemblies that use electro-magnets are a lot easier to work with because you can turn off the torque by turning off the power to the electro-magnet. This makes the mechanical design a lot simpler. Now you do need electrical power and control circuits in order to get them to operate. Because of this, the designs using this new technology will always be more powerful motor.

Most of my motor designs in my books use electro-magnets in them. It is how I use them that is different than in the motors we use today. The designs are in theory form, but if they can be used in cars to double the range of cars, even that little bit of improvement can make a big difference in the world. I believe my designs will make a much bigger difference than that.

As my motor designs increase the torque they produce with a reduced power input to them, there has to a point where the designs will cross the threshold into a free energy device.

The top two lines on the drawing show the function of the switches in the stator assembly that are used in generating the traveling flux wave in the stator assembly. The following three lines on the drawing show the forces upon the switching wheel assemblies themselves. These switching wheels to my knowledge have not been used in motor designs before.

Now for the big question? How does the continuous torque the rotor has to turn the rotor compare to the switching wheel losses? If the losses are less than the rotor forward torque, then the rotor will have continuous movement. This is before I add the electro-magnets to the rotor assembly. Since creating a functioning motor on permanent magnets alone is low, building the motor with the alternating electro-magnets into the motor is a must for the proto-type design. Now with that being said I will be looking to the electro-magnets being fed from an on-board generator. The on-board approach will allow me to reduce the hardware and circuitry than the conventional way of having a separate motor and generate to operation in motor in a test system. IF and that is IF the electrical pulses the generator generates on the on-board generator is enough power to power the electro-magnets in the rotor to create continuous movement, then I would have a COP>1. This is why I am designing this into the proto-type before building the motor. Even if the on-board generator works on this motor design, the motor would have to have a way to bring the motor up to speed. One option is to start the motor with another motor and once operational, then the external motor could be removed from the circuit. Now another option would be to drive the electro-magnets from an external power supply until the motor comes up to speed and then it could be removed.

Now if this still is not enough to operate the motor, I will still have the option to drive the electro-magnets with a specially designed external supply. Then an evaluation of the efficiencies would be required to evaluate the motors performance. The external power and control can use the following modified TANK circuit for it. This circuit is an improved circuit bringing another new technology to the motor world of design.

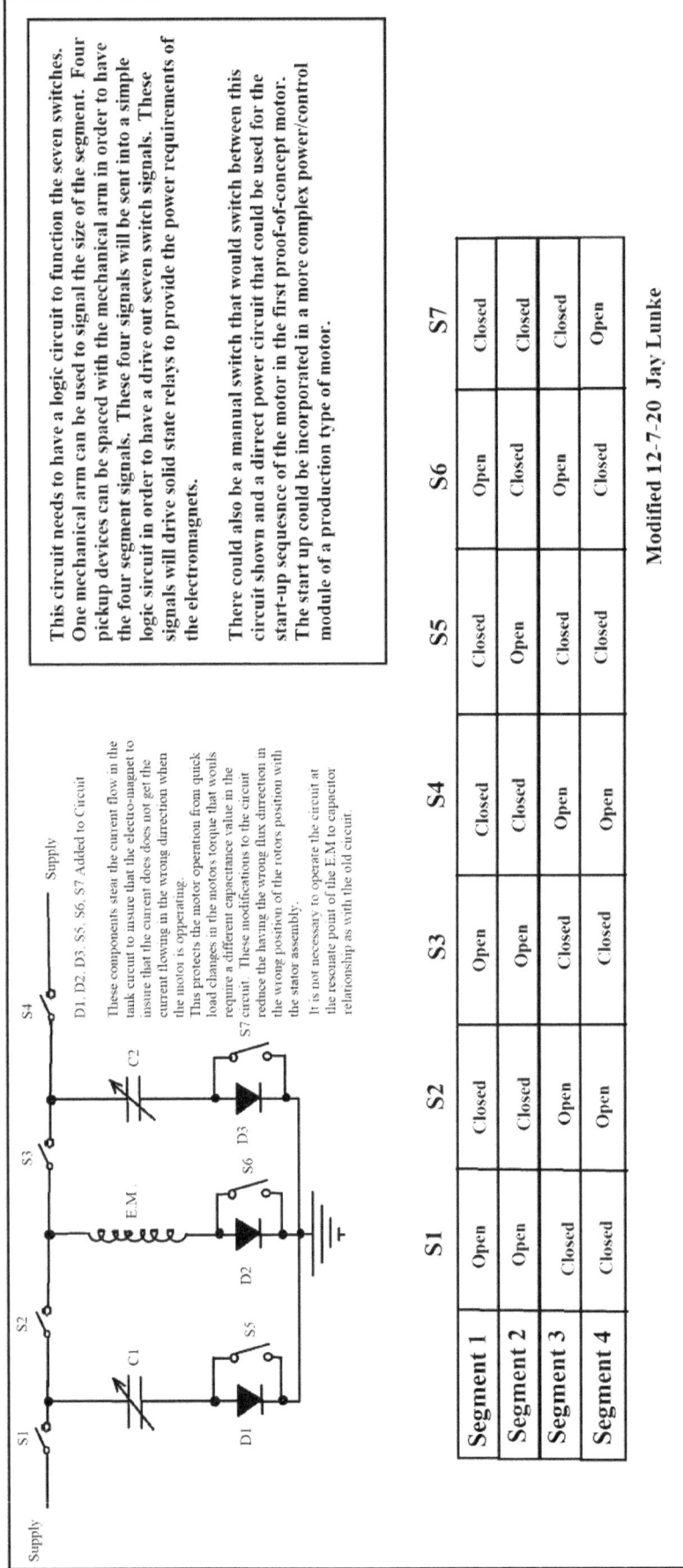

This circuit needs to have a logic circuit to function the seven switches. One mechanical arm can be used to signal the size of the segment. Four pickup devices can be spaced with the mechanical arm in order to have the four segment signals. These four signals will be sent into a simple logic sircuit in order to have a drive out seven switch signals. These signals will drive solid state relays to provide the power requirements of the electromagnets.

There could also be a manual switch that would switch between this circuit shown and a dirrect power circuit that could be used for the start-up sequesnce of the motor in the first proof-of-concept motor. The start up could be incorporated in a more complex power/control module of a production type of motor.

D1, D2, D3, S5, S6, S7 Added to Circuit

These components steer the current flow in the tank circuit to insure that the electro-magnet to insure that the current does does not get the current flowing in the wrong durrection when the motor is opperating.
Thus protects the motor operation from quick load changes in the motors torque that would require a different capacitance value in the S7 circuit. These modifications to the circuit reduce the having the wrong flux durrection in the wrong position of the rotors position with the stator assembly.
It is not necessary to operate the circuit at the resonate point of the E.M to capacitor relationship as with the old circuit.

	S1	S2	S3	S4	S5	S6	S7
Segment 1	Open	Closed	Open	Closed	Closed	Open	Closed
Segment 2	Open	Closed	Open	Closed	Open	Closed	Closed
Segment 3	Closed	Open	Closed	Open	Closed	Open	Closed
Segment 4	Closed	Open	Closed	Open	Closed	Closed	Open

Modified 12-7-20 Jay Lunke

This modified tank circuit can be used for several other exsisting motor designs. This book spends a lot of time describing the tank circuit operations without the three diode and those three switches across them in a later chapter. The added diodes and switches allows more slop in the capacitance value for the operation of the circuit. This will allow a simpler control circuit because the old design needs to operate at the resonent point of the capacitor and electromagnet. In the un-modified circuit, each time the load and/or speed of the motor changes, the capacitor needs to be adjusted for those changes. Now using this power scheme with this motor design has a good chance of having a COP>1.

Now if it gets to that point, then this would not be the motor design that anyone would be interested in developing because the cost to build this motor design would be too expensive. They would use the new modified TANK circuit with a different motor design.

Now there may be a way to perform of moving the switching magnet into and out of the area that the rotor rotates passes around the stator assembly perpendicular to it during the operation of the motor assembly.

The following drawings are sone optional ways for the switching to occur. The are for sure many other ways to do it as well. The switching magnet being a permanent magnet, will not work as long as the resistance to the movement of this magnet is more than the torque produced in the rotor to rotate in the motor assembly.

All Permanent Magnet Motor

This is a modification of the all permanent magnet motor I built that is redesigned to reduce the resistance in the switching wheels rotation. It also changes the orientation of the stator permanent magnets to increase the torque between the rotor and stator assemblies. The old switching wheels had four magnets, two in the posative direction and two in the negative direction. The new wheel has one magnet followed by an open space. This configuration will create eight traveling flux waves in the stator assembly.

These two changes will reduce the switching resistance at the same time as produce forward torque between the rotor and stator assemblies. The big question is will the torque generated in the rotor in this motor design be greater than the resistance to the turning switching wheels. This is something I do not know how to model. So if you have modeling software, could you model this motor design?

You need to go to the drawings already posted about this motor design to see how the rotor permanent magnets can rotate purpandicular to the stator assembly that has the stationary permanent magnets along with the switching wheels. From the lessons learned, I would change to much larger shafts and reduce the number of gears from 17 down to 9 gears.

The big reason I believe this design needs to be built and tested is because this motor uses the Three Layer technology that keeps permanent magnets between the rotor and stator assembly from having a sticky point. It does this by constantly reconfiguring the stator assembly during the rotation of the rotor in the motor. The switching wheels perform the reconfiguration need to achieve that objective.

With the switching wheel now has the orientation of the magnets in the same direction, the resistance will be a lot less than the older motor design. The attraction of going into the switching postion with the repultion of leaving the switching position should for the most part cancel each other out.

modified 12-7-20 Jay Lunke

Rotor

Switching Wheel | Switching wheel | Switching Wheel | Switching Wheel

Stator Assembly

Rotor

Switching Wheel | Switching wheel | Switching Wheel | Switching Wheel

Stator Assembly

8:1 Rotor to switching wheel ratio

Soft Switching Option of Three Layer Technology

Stator Permanent Magnet

Sliding Switching Magnet

Slide Assembly for Permanent Magnet

Arm, Part of cam assembly to move switching magnet on slide assembly

The Rotor assembly is not shown. It has magnets that expand over one stationary and one switching magnet. THis is followed by a gap the same size. This repeats around the full rotor. The rotor runs perpandicular to the stator assembly at the diameter were the switching magnets are fully extended. The switching magnets are designed to remain on the sides of the permanent stationary magnets in order to reduce the ingauging and disingauging forces that reduce the power of the overall motor performance. The rotor to stator are interacting with each other to always have forward torque on them. This design is an attempt to reduce the resistance to the switching action while the motor is in operation.

Each of the switching magnets are connected to a cam system that is in sync with the rotor to create the flux waves that the rotor rides on. I do not show the cam hardware in this drawing. The rotor could be build in a drum with another one of these stator assemblies build on the other side of the rotor drum. THe stator and switching magnets on the other drum would need to be the other polarity for the magnetes.

modified 12-7-20 Jay Lunke

Now if you can come up with a way of performing the switching movement with less resistance than the torque created in the rotor assembly, you will have a great over unity motor on your hands.

The last two chapters were added before the following chapters. These chapters are important in understanding the new technology and also add several options for difference motor and power circuit designs that can be used in motor design.

WHY DO I BELIEVE THAT THIS MOTOR MEETS OVER UNITY CRITERIA?

There are permanent magnet motors that use 50% of the electrical energy by using the same number of permanent magnets as electro-magnets in the design. The best motor design I have come up with uses 25% of the energy over the conventional electric motors. The way I have been able to do this is to build the motor with my Three Layer Design for Electrical Mechanical Movement. By using this new technology in a motor design, I will have an efficiency rating of the motor functioning at 50% of the permanent magnet motor. With efficiencies of 25% over conventional electric motors and 50% over magnetic motors, I would be in ranges that some people would label as over unity or free energy devices. I will show you why I stay within the frame of meeting the laws of physics in order to accomplish this.

In addition to the improvement to the motor design itself, I have designed an electrical drive circuit for the motor that captures and reuses part of the electrical energy used in the electro-magnet that creates the magnetic field in the motor. This circuit along with the motor design will create one of the best motor systems ever created. I want to give these plans free to the rest of the world.

WHY OVER UNITY BY ITSELF MAY NOT BE THE BEST WAY TO GO

There are other energy sources that may make more sense to use than an over unity device. I have 27 solar panels on my roof that are creating electrical energy for my house. The cost of the over unity motor and generator may cost more than the solar system I have on my roof. If that is the case, then I will choose the solar cells for my energy needs for my home. There are other technologies like wind generators that may be better options than some of the over unity devices in the world today.

WAM, BAM, ANOTHER SCAM

There are so many scams and false information out there, that it is hard to know what is legitimate and what is false. I have been fooled for a while until I have done enough additional research to reach a conclusion about things. When people are asking for money to further develop the technology, a red flag should go up. Now it may be legitimate, but you should do a lot of research about them before giving them money. But I truly feel sorry for the people who are legitimate, because when there are so many scams out there for people's money that many people will believe that everything is a scam. I like to look at both side of information by looking at many sources about it before coming to a conclusion. I like to read articles about scams people have found, because it makes me more knowledgeable about them.

Why over unity motors will not change the world over night

It takes time to design and manufacture motors. The development costs of products are usually incorporated into the cost of the product when it hits the market for a few years. This means the upper-class people will get them first. But as the market begins to change, it will reduce the cost of the other energy sources to keep people using the old technologies.

The greatest fear I have with this new technology is that of the theft of the motors will be a large problem. There may need to be a legislation to register the motors with a unique number stamped into each motor.

I personally would like to see people use the savings that they realize from the motors to help people who are less fortunate than they are.

THREE LAYER DESIGN FOR ELECTRICAL MECHANICAL MOVEMENT DOCUMENT

Invention Objective:

The overall objective of my research is to revolutionize the way mechanical movement is done using this new three-layer technology. These are three layers of function not hardware. Although many times the hardware is in three distinct layers. There is a stationary assembly I call the stator and there is the moving assembly I call the rotor. Now one of the outer layers are built up using permanent magnets into the rotor assembly. The other outer layer is built up using permanent magnets in the stator assembly. The middle layer can be built in either the stator or the rotor assembly. For the rest of the paper, I will assume it is in the stator assembly unless I write otherwise. The middle layer is unique to other motors in that it changes the way the permanent magnets in layer one respond to the permanent magnets in layer three.

OVERVIEW WRITTEN DESCRIPTION OF THE THREE-LAYER TECHNOLOGY:

This new technology is divided up into segments of travel. Each segment is about the length of one permanent magnet. The mechanical movement produced by the technology is built up of segments of two or four segment sets of travel where the segment sets repeat themselves over and over again into either linear or circular movement in the mechanical device. Layer 1 and 2 make up the stator. Layer 3 makes up the rotor. The following shows the two-segment set configuration.

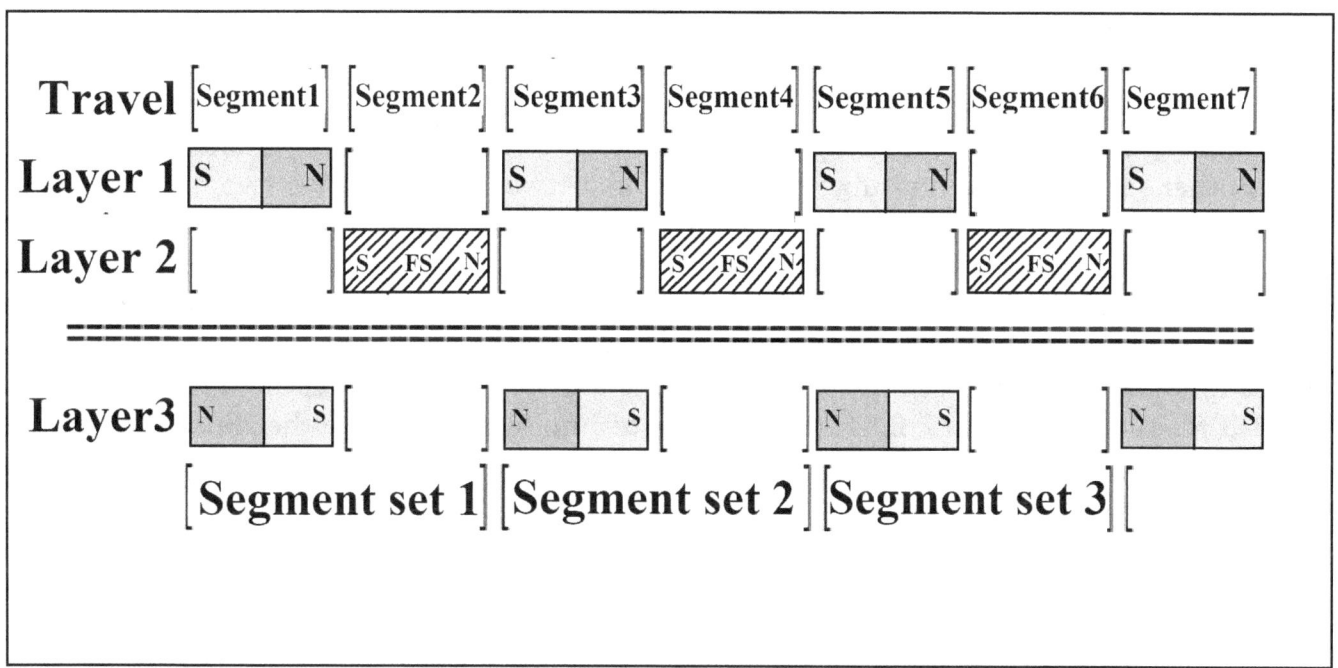

The two-segment component configuration is built up of three layers. Layer one has a magnet in segment one and nothing in segment two. Layer two has nothing in segment one and a flux switch in segment two. Layer three (the rotor layer) has a magnet in the segment followed by no components in the next segment. The two-segment set repeats itself over and over again. Since the rotor moves in relationship with the stator, then I talk about the rotor magnet as it comes into play with each segment of travel.

For the two-segment rotor travel, the first segment of travel is produced by the magnetic attraction of one stator magnet and one rotor magnet to each other to pull the two magnets as close to each other as the mechanical device allows. This action occurs with the two outside layers of the technology.

Description of segment 1
Stator Layer 1; magnet_____ [S PM N]
Stator Layer 2; no components in this layer for segment 1
Small air gap between stator and rotor assemblies
Rotor Layer 3; mag. causes layer 3 to align with layer 1's PMs____ [N PM S]

In moving through the second segment of travel, the flux switches in the middle of the three layers of technology are activated producing an easier route for the flux lines from the adjacent stator magnets to flow through the flux switch. The flux switch can be either a mechanical or electrical device. When enough energy is supplied to an electrical flux switch, then the flux switch not only has enough flux to change the stator flux flow through itself, but it will also have enough flux flow to attract the rotor's permanent magnet. The rotor magnet aligns itself with the flux switch. So, the flux switch changes the flux flow of the permanent magnets changing the dynamics of magnetic forces between the rotor and stator to attraction moving the rotor one segment of travel.

Description of segment 2
Stator Layer 1; no functional components in this layer, in ring magnet configuration
Stator Layer 2; Flux switch uniting with adjacent stator magnets___ [S EM N]
Small air gap between stator and rotor assemblies
Rotor Layer 3; magnet causes layer 3 to align with layer_____ [N PM S]

The Four segment component configuration is built up of three layers. Layer one has a magnet in every other segment with nothing in the segments between them. Layer two has nothing in segment one and three. A flux switch is in segment two and four. Layer three (the rotor layer) has only one magnet in the four segments. The other three segments are blank. When other segment sets are added to the device, then the rotors need to line up so that the magnets are in every forth segment. Since the rotor moves in relationship with the stator. I talk about the rotor magnet as it comes into play with each segment of travel. The four segments are repeated multiple times as needed for the final motor configuration. This technology supports either linear or circular motion. In circular motion the additional segment sets will come around to the beginning of the assembly to form either a ring or disk-shaped device like a motor.

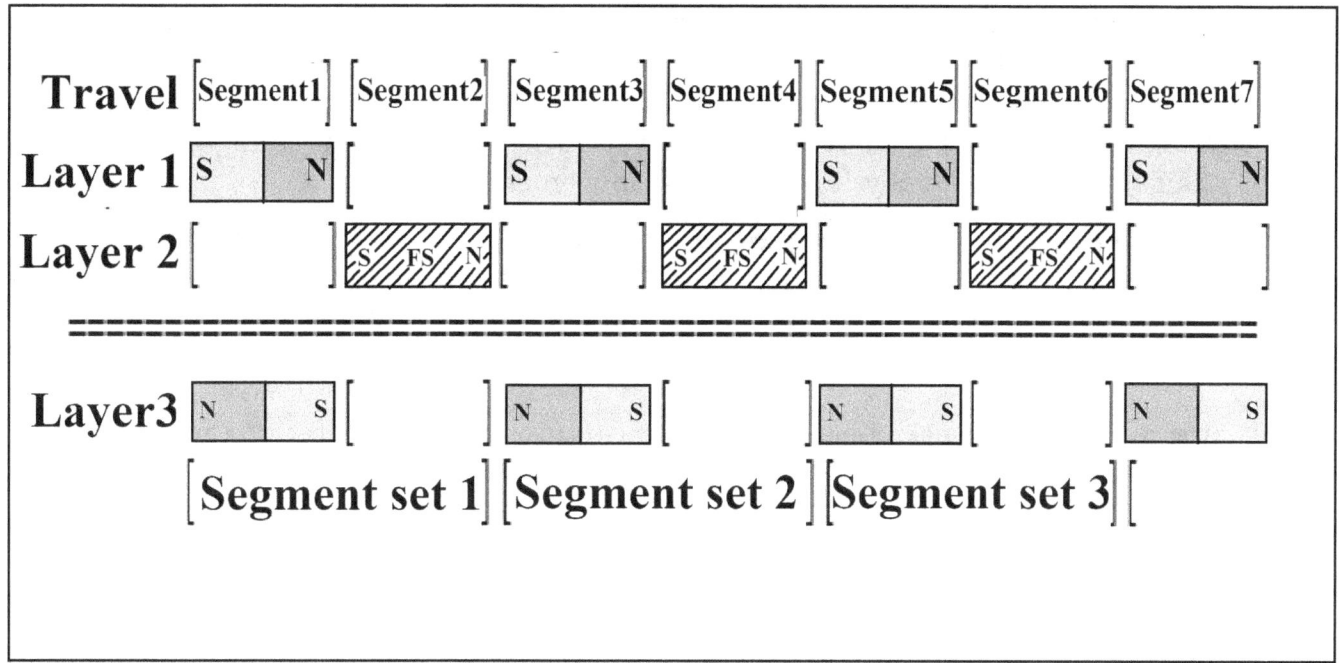

For the four segments of travel, every time the segment of travel has a magnet in it, then the travel is produced by the magnetic attraction of one stator magnet and one rotor magnet to each other to both push and pull the rotor magnet to the end of the segment travel. Notice that I rotated the permanent magnet in the rotor so that one pole of the rotor magnet interacts with the stator magnet. The reason for this is because the rotor magnets perform better with larger functional stator magnets in this configuration.

With the four-segment configuration, not only are the number of rotor magnets are cut in half, but the flux switches are activated alternately every other flux switch in the motor in alternating segments of travel. This means that each individual Flux switch is activated in every fourth segment or 25% of the time. When a Flux switch is activated, then the adjacent flux switches are not active.

Let's look at the above configuration at the movement of PM1 in layer 3 as it moves from segment 3 into segment 4. FS2 in layer two is activated. FS1 and FS3 are not activated. With the activation of FS2, the flux flow from PM2 and PM3 go through FS2 creating a new functional magnet three times bigger than the individual magnets. The stator magnets do not form a ring magnet with the other flux switches and stator magnets like the two-segment configuration because the alternating flux switches are turned off. The rotor magnet now has to interact with the functional magnet in the stator assembly that causes the rotor to move one more segment of travel. The re-routing of the flux lines is what causes the attraction in this segment to have forward torque in it. Then all the flux switches are deactivated. This action starts to move the rotor through another segment of travel. This segment is another rotor to stator magnet interaction to move it through the segment. The following segment movement will now require the adjacent flux switches to be activated for another segment of movement to occur. FS1 is powered off now during this segment of travel. The torque will not be as great as the two-segment configuration because the torque is at the optimal point when the stator magnets are shorter in length and the rotor and stator magnets are the same size.

When using an electro-magnet for the flux switch, you only need ½ of the electrical energy of the two-segment configuration to power the electro-magnets to create attraction and movement in the segment of travel. The four-segment configuration uses a 25% duty cycle on the individual flux switches to create the power to move the rotor through the segment set of travel. The four-segment configuration uses one quarter of the power of the two-segment configuration having more than half of the power per physical size of that movement.

The three-layer technology can be designed to be used twice around the rotor to increase the motor's power as is shown in the diagram below. When doing this you would want to have stronger rotor magnets so you do not chock the potential power in the motor assembly.

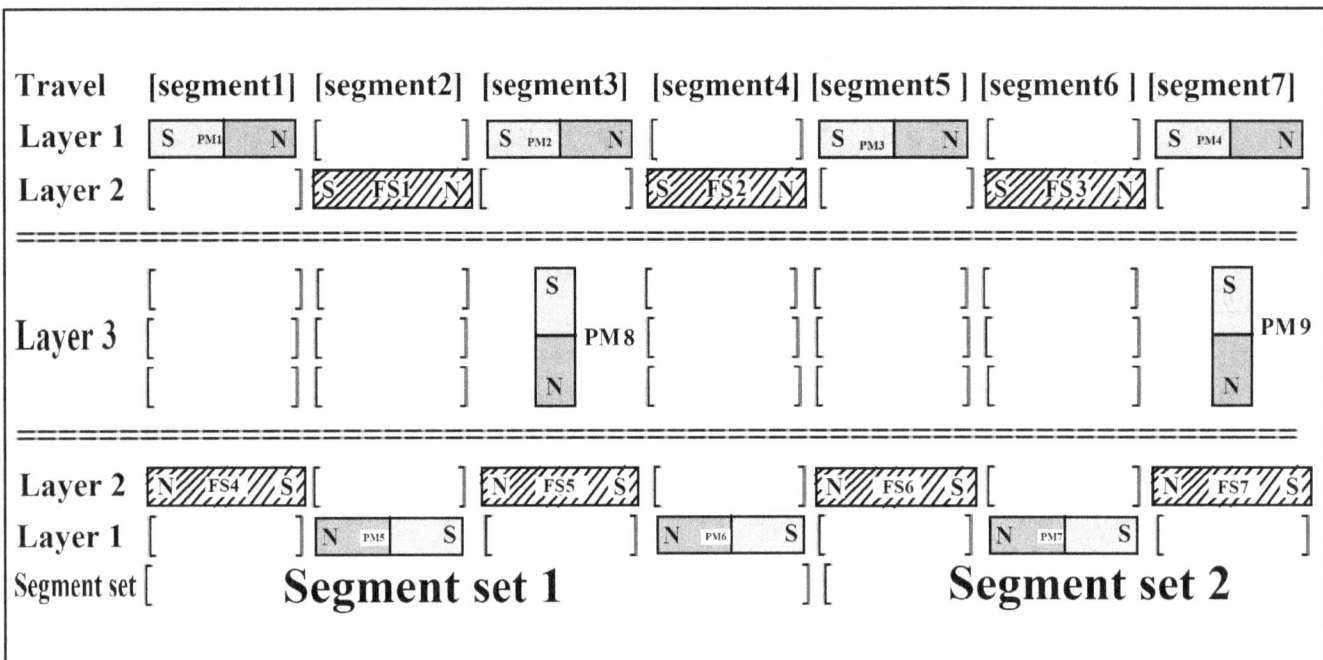

Designing a segment offset between the stators on each side of the rotor will provide a smoother operation of the motor. It does need more switching. When you build a device with two three-layer technologies on each side of the rotor assembly, you want to offset layer 1 on the bottom with layer 1 on the top by one segment. You also want to offset the bottom layer 2 with the top layer 2 by one segment. The flux switches in the bottom layer 2 will alternate switching with the switching in the top layer 2. The individual switches will be operating at a 25% duty cycle. There will be one flux switch active in each segment of travel. In each segment of travel there will be one rotor magnet to stator magnet interaction of attraction to each other. This design configuration will provide a very smooth, powerful and efficient electrical-mechanical device.

One way to look at this new technology is as follows:

There are two solar panels on a roof, one of the panels is called attraction and the other one is called repulsion. There is a third component which is a mirror with a spring attached to it. The sun hitting the solar panels create the same amount of energy in the panels over the course of the day. Now let's say a person standing on the roof pushes the mirror six inches for one minute, let's go for one minute and then repeats the process all day long. Now when the mirror is left along, the panels have the same amount of light on them producing the same amount of energy. When the mirror is pushed six inches, it takes the sun light that normally would fall on solar panel one and directs it onto panel two. At the end of the day panel, one got ½ of the light it normally would receive while panel two received 150% of the light that it normally would get because of the switching of the mirror. So, no energy was lost in the system because 50% and 150% is equal to the 200% of energy in the two panels that would be produced without the mirror.

The three-layer technology does not change the amount of flux flowing from the permanent magnets. It uses it directly, then indirectly through the flux switches when in the correct position of the rotor assembly in order to optimize the forward torque of the motor assembly. The three-layer technology is very simple to use. It is a building block for unlimited electro-mechanical movements with efficiencies not seen in electro-mechanical devices before.

MORE ABOUT THE THREE-LAYER TECHNOLOGY:

The three-layer technology optimizes the use the magnetic forces of permanent magnets into creating mechanical movement. Without the flux switching in layer two, the rotor permanent magnet resists movement from the stator permanent magnets stopping the rotation. As segment one and two alternate having attraction and repulsion forces between each other, these forces cancel each other out.

The movement of the rotor is through the attraction of the rotor magnet to the stator magnet is segment one. The movement of repulsion is in segment two travel. The travel of segment two without the flux switching is the opposite direction as segment one. So, the key of the three-layer technology is to use the middle layer of the technology into creating a torque in the same direction as that of segment one. Since moving either the rotor or stator magnet out of their permanent position during this segment of travel could require as much or more energy to move them, I decided not to spend a lot of time using that direction of design for the rest of this book. This is where a third level of components are introduced to the design. People have tried to shield the flux of repulsion during the second segment of travel. I did not see this as an optimal design solution because there remains a lot of untapped magnetic force not taken advantage of with that approach. Since magnets are always looking for the easiest path for the flux to travel in to complete the flux flow of the magnet, I decided that this would be the key to the new technology. Since the rotor and stator magnets are aligned with each other at the end of the first segment of travel, the flux switches need to provide an easier path to flow than that of the interaction between the rotor and stator magnets. It is easier to change the stator configuration than the rotor assembly and this is why the flux switches are in the stator assembly. The rotor and stator need a little distance between them in order to allow the flux switch to take control of the flux lines. In the flowing example, the flux switch is made of copper wire on an air core so that when the power is off, the permanent magnet's flux lines cannot see them. Layers one and two are over-lapped 100% in the example.

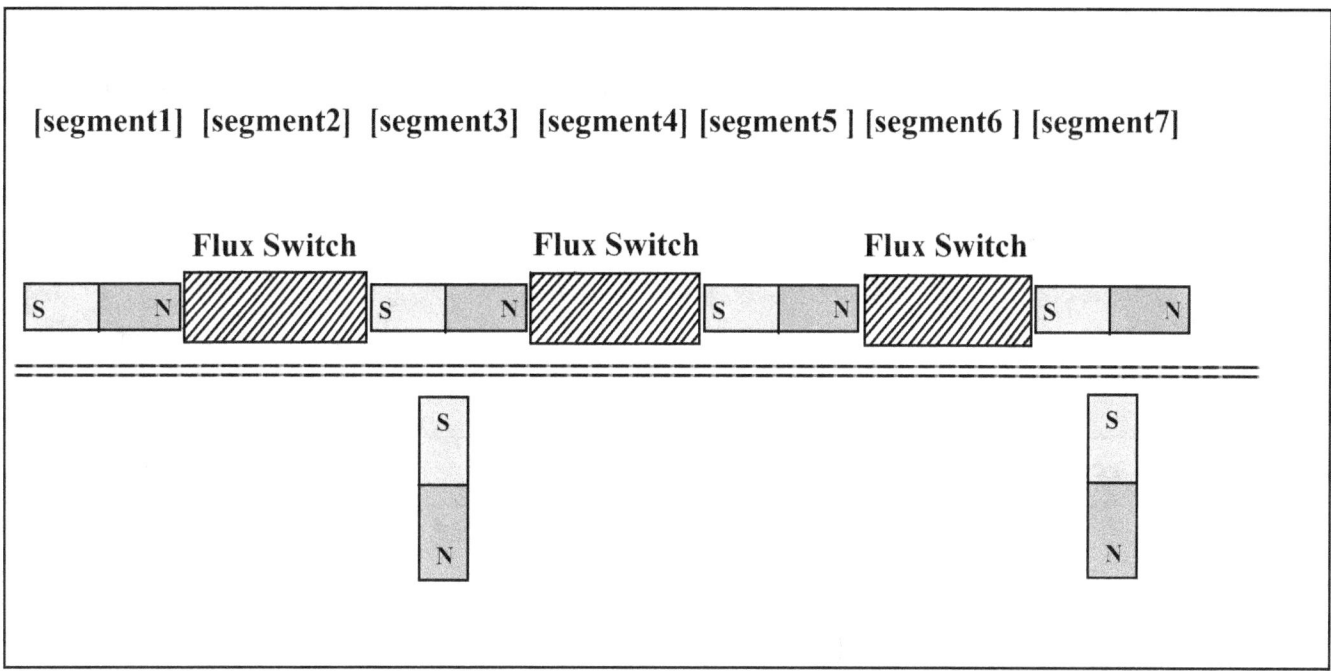

When the string of stator magnets and flux switches stretch around into a circle, then a ring magnet is formed. The flux from each stator magnet and flux switch is self-contained in the circular movement of this ring magnet. Additional flux in the electro-magnets interact with the rotor magnets in order to move the rotor through the second segment of travel. The flux switch can be either electrical or mechanical in nature. When electrical, the electrical energy goes through a coil to create the magnetic force of flux in it. The flux flows through the adjacent stator magnets. If it is a mechanical switch, then a magnet or flux carrying material is moved into or out of the position between the stator magnets to provide the easier path for the flux to flow through.

With overlapping layers 1 and 2 and using the three-layer technology on both sides of the rotor, using the four-segment configuration is a simple, manufacturable motor assembly to build. With the rotor built into a disc assembly and attached to a shaft could produce a very competitive product.

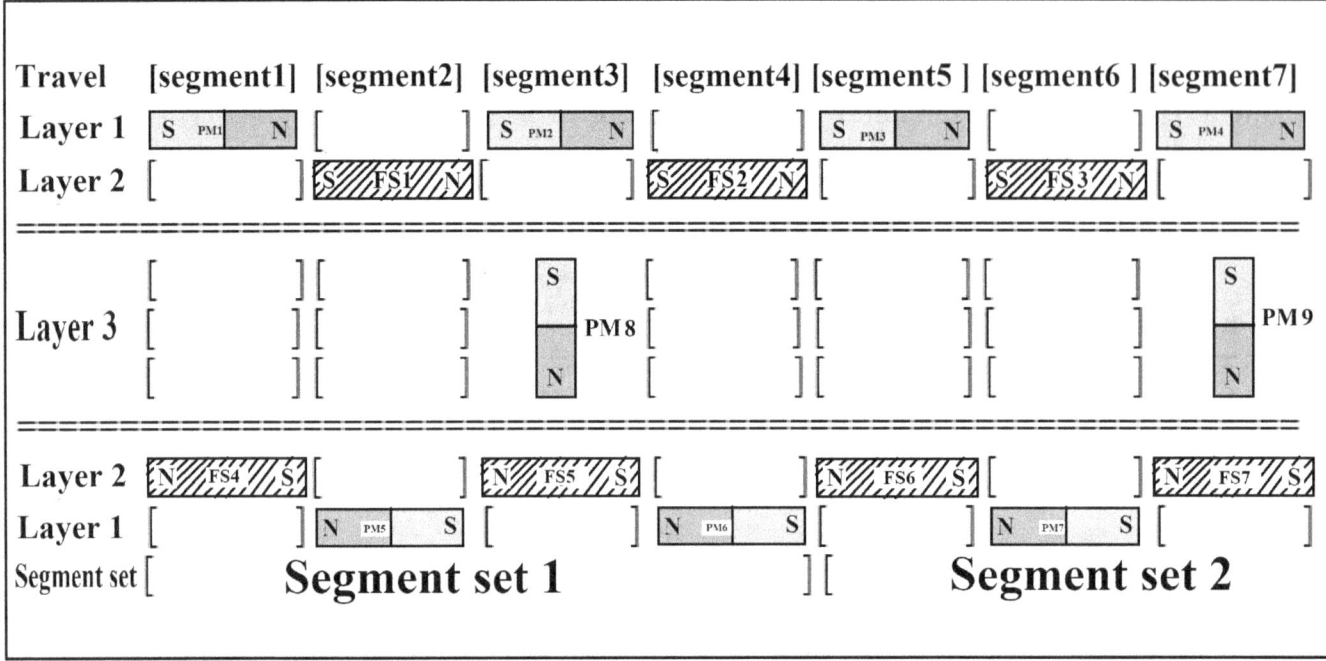

The movement of the mirror with the spring in it has an attraction and repulsion force of moving it. These two forces cancel each other out. The mechanical flux switching in the three-layer technology has attraction and repulsion forces for moving the switch in and out of the flux switching position that cancel each other out. So, the largest losses in the well-designed mechanical switching are frictional losses.

When electrical energy is used to perform the flux switching, then there are ways to recapture some of that energy like the spring in the mirror. I will be getting into detail in how this works later on in the paper.

The three-layer technology should revolutionize automated mechanical movement in the way object-oriented programming has been changing the way software programming is done in computers today. My goal is to have an energy efficient alternative for many of the mechanical movements in electro-mechanical devices that are built today sometime in the future. It would include many of the devices that produce mechanical movement. The new technology will produce mechanical movement using less electrical power than current electro-mechanical devices.

When no electrical energy is sent to the motor device using electro-magnet flux switching, then it will be in a soft locked condition by the permanent magnets that operate in it. This condition can be used to the advantage of the designer when the power is off in these devices.

When the mechanical device is a motor. When the flux switching is mechanical it needs to be moved in and out of its flux switching position. You want to do this fast at the same time limit the losses in that mechanical system to do that. When you achieve this, then the motors would need to have the breaks on to stop it. You could design some other ways to stop it. The usage of these motors would most likely be restricted to industrial applications where safety controls can be put in place to keep people from getting hurt. As an option, you could consider designing the motors with a rotor disc that can be pulled out of reach of the stator magnets to stop the motor.

In reality, I do not see being able to move these switches in and out fast enough without creating large frictional losses to the system until now. So, you would have some repulsion forces that would occur that would create torque in the opposite direction from segment one's movement. That plus the friction losses of moving the flux switches in and out of position would hurt most mechanical switching designs. For the mechanical flux switches to work they will likely be in slower rotating motors. So, there are applications for motors with both mechanical and electro-mechanical switching in them.

Definitions:

I will use naming conventions that are used on common day electric motors. I will use these terms whether I am talking about one of my motors that rotates in a circular motion or one that operates in linear motion. The stator will normally refer to the stationary part of the motor and the rotor will normally refer to the moving part of the motor assembly.

Stator = I use this term to identify the motor sub-assembly that has two magnetic layers in it. In most applications, it will be the stationary part of mechanical device. The two layers of magnets will consist of functional flux switches in one layer and permanent magnets in the second layer. The function of the stator is to interact in two different ways. The first way is one having the flux switches inactive followed by the second way of having the flux switches active. When the flux switches are active it changes the interactions of the permanent magnets in both the stator and rotor assemblies to produce positive movement in the motor device. This is done in two different ways, depending on the motor design, that will be described later in this paper.

Rotor = I use this term to indicate the assembly in the electro-mechanical device made up of one layer of permanent magnets. Usually, the rotor assembly is the one that moves. The moving mechanical assembly will interact with the magnets in the stator assembly to create the physical movement of the device.

Note: Either the rotor or the stator can be the stationary assembly in this technology, but then the other mating assembly will need to be the moving assembly. Besides the technology being used in circular movements in many applications, the technology may have linear movement in applications like mass transit trainset systems.

Sub-assembly: Several physical components installed together to create an assembly like a disk, rotor or housing assembly.

Functional Sub-assembly: select components of two or more sub-assemblies that perform the function of movement between the rotor and stator of one segment.

P/P = It is the functional sub-assembly built with physical components, usually one permanent magnet in the stator assembly and one permanent magnet in the rotor assembly. These magnets interact with each other to move the rotor in relationship to the stator a distance of one segment. This interaction is the power of the flux lines between the two magnets. When the magnets are built into the parallel configuration, then the north pole of the rotor magnet pulls the south pole of the stator magnet as close as possible to each other. At the same time the south pole of the rotor magnet pulls the north pole of the stator magnet as close as possible to each other.

Note of caution when designing the P/P and F/P devices, if the permanent magnets are too close to each other in either the rotor or stator assemblies, then the magnetic lines of force would move through the adjacent magnets instead of interacting with each other between the stator and rotor assemblies.

F/P = It is the functional sub-assembly made up of components used to create the physical movement of one segment of movement when the flux switch is applied to it. When the flux switches are in activation mode, it causing the magnetic forces a change their behavior in this functional sub-assembly. These changes cause physical movement between the rotor and stator assemblies of one segment of travel. This is usually referred to as the second segment of motor movement.

Segment = It is the physical movement in the P/P or the F/P functional sub-assemblies caused by the magnetic field interactions in them. Note: The physical movement is from the start to the end of the segment.

Segment Set = It is the physical range of movement in the functional sub-assemblies. It is equal to one permanent magnet and the following empty segments until another permanent magnet comes alone. There

PFMMD = Permanent / Flux Switch-Permanent Magnet Movement Device; These are electro-mechanical devices build up using both the P/P (magnet to magnet interaction) and F/P (Flux switch to permanent magnet interaction) mechanical assemblies in them to create the "segment set" movement through them. Many applications are built with multiple PFPMMDs in them to create motors and other electro-mechanical devices.

Motor = I use the term to mean any complete sum of electrical mechanical P/P and F/P devices using the three-layered technology to complete its operational function.

Electro-magnets = This is a coil of wire wound around a material that is not attracted by the permanent magnets in the motor assembly. I need a material that carries the most flux lines when the electricity of the electro-magnet is turned on without attracting flux lines when the power to the coil is off. It may need to be an air gap.

Flux switch = This is a device that controls flux lines by redirecting them through itself and the adjacent permanent magnets adjacent to it. It is either electro-magnets, permanent magnets or flux carrying materials having low permeability.

THREE-LAYER TECHNOLOGY OPTIONAL CONFIGURATIONS:

The following arrangements of the permanent magnets are common. I also do not show the distinction of the rotor and stator assemblies. One magnet will be in the stator and one in the rotor assembly. These are only shown here to show that I start with common configurations and then introduce the three layers design to them that makes them unique from other electro-mechanical designs. In order not to confuse the reader at this time, I will not be showing the flux switches in the following examples. The flux switches can be installed in many different places in the device as long as two things happen.

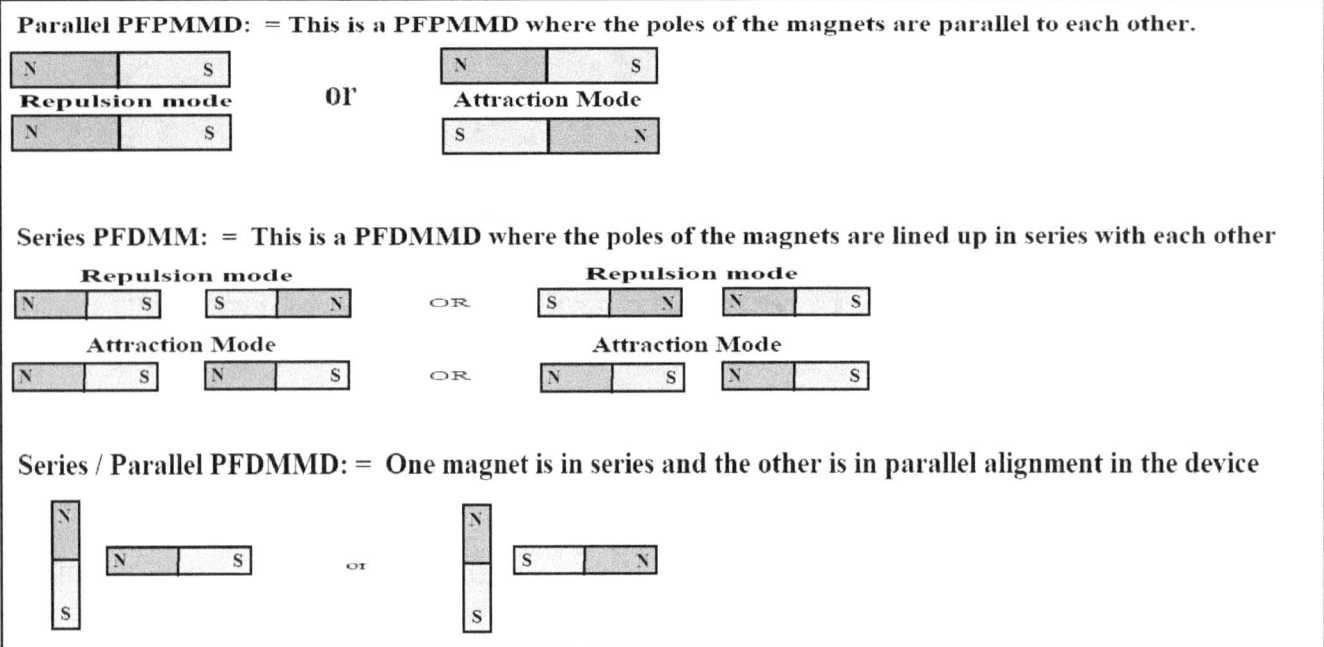

Series/Parallel PFPMMD: = One magnet is in series and one magnet is in parallel alignment in the device. I use the Series/Parallel configuration in my efficient rotor motor designs where the end of the segment movement is in the attraction mode. The functional stator magnet has positive torque through the full segment travel with the power of the electro-magnet only producing the same flux lines of the permanent magnets in the functional magnet.

MMD = Magnet Movement Device

PFPMMD = Permanent / Flux Switch-Permanent Magnet Movement Device; (2 segment movement with the first being only permanent magnet to permanent magnet interaction. The second segment movement being a flux switch to permanent magnet movement.)

There are several PFPMMD device designs, some producing greater efficiencies than others. Some devices have better manufacturability while others are more suited for special applications.

PFPMMDA = Permanent / Electro-Permanent Magnet Movement Device Assembly = When more than one PFPMMD device are connected together to create an assembly.

In most of the assemblies the assemblies are built up from the same type of PFPMMD but that is not necessary. Also, one or more PFPMMD components can be shared with other PFPMMDs.

DISK PFPMMDA:

The disk PFPMMDA can be built using either the series/parallel or Parallel PFPMMD's that are in either attraction mode or repulsion mode. I personally prefer the attraction mode in order to maximize the life of the permanent magnets and the benefits of the flux switching. The disk designs function very similarly to the linear movements that could be used in a mass transit system. The nice thing about the disk assemblies is that they are a lot smaller than the mass transit assembly. The disk assemblies can be used for designs like electric motor assemblies.

Note: The magnetics are not shown to scale in this paper. The magnets in the stator are closer to the magnets in the rotor than they are to each other. This also means that the magnet in the rotors is closer to the stator magnets than to each other.

Getting a closer look at how the technology works.

New Design Uniqueness (three-layer system)

What is unique with this new technology is that it adds an additional functional layer of flux switches between the other two layers of magnets. When the flux switches are activated, they change the way the magnets between the rotor and stator assemblies interact with each other. The three layers in this design is what is unique when compared to other magnetic, electric and magnetic\electric device designs. The middle layer consists of flux switches that switches on and off again during the movement of the motor assembly. The flux switches turning on and off again does not change the amount of flux flowing through any of the permanent magnets in the motor assembly. What it does is to changes the route of flux flow for the permanent magnets. The flux switches change not only the interaction of the permanent magnets between the rotor to stator assembly, but also between the permanent magnets in the stator assembly to each other as well. Both of these changes create more positive torque for the motor's movement. It does this in a way that creates more efficiency of the motor assembly.

The physical build of the motor assembly is built in three layers.
Layer 1 is normally built with permanent magnets. They can be substituted with electro-magnets.
Layer 2 is built with Flux Switches.
Layer 3 is normally built with permanent magnets. They can be substituted with electro-magnets.

In most motor designs, layer three is built into the rotor assembly and Layers one and two are built in the stator assembly. But some designs are built with layers 2 and 3 in the rotor assembly while layer 1 is in the stator assembly. All three layers can be physically built on three different assemblies, but the layer two needs to be either stationary with either layer 1 or layer 3 of the motor assembly.

When describing the physical assemblies, it becomes hard to describe the operation of this technology because some of the mechanical components have dual functions in the motors operational characteristic's. Also, components from three different physical assemblies are used to create one function through one segment of the motor's physical movement. Because of this I will be describing the technology mostly in terms of functional assemblies. These functional assemblies are what creates the physical movement in the motor. The mechanical movement is broken up into two functional sub-assembly, the first one I will call part P/P. The second one I will call part F/P. The segment that I will be talking about refers to the physical movement that takes place in the device during the operation of one of these functional assemblies. Most of my mechanical devices have a minimum of one P/P functional sub-assembly and one F/P functional sub-assembly. Each functional sub-assembly creates one segment of movement. When using the power rotor assembly these two functional sub-assemblies together create a two-segment movement I call the "segment set".

The efficiency rotor has one magnet for every fourth segment. With that being said, the function of the motor is derived from the action of the rotor magnet to the stator components. So, a segment set occurs in two segments of travel and then it repeats itself. But when it repeats itself, a different stator permanent magnet and functional magnet interact with that same rotor magnet. So, it takes two sets of "magnet sets" for the usage of the one rotor magnet. For the efficiency rotor, I often refer to four segments of rotor travel for one segment set". Note that some of the same physical components can be used in both the P/P and F/P functional movements in the technology. What makes up each functional sub-assembly is the active components making the movement in each of these segments. These two functional sub-assemblies together I call the PFPMMD (Permanent / Flux Switch-Permanent Magnet Movement Device). Now several of my mechanical assemblies have multiple "segment sets" in them. These multiple sets will either be configured to travel in a linear line or they will be arranged curved so that when enough of them are put together they intersect the first segment again to form a circular, ring or disk type of mechanical assembly. This new technology is good for designing many efficient alternative mechanical devices. They can be used in place of many of the current electro-mechanical devices like electrical motors used in today's products. Since my motor will be more expensive to build using this new technology, the old technology will always be with us. Mostly in the very small applications where efficiency is not as much of a factor to the consumer.

Segment 1, P/P functional assembly

What happens in segment P/P to create the part P/P segment movement? First of all, I want to say that segment part P/P would not be possible without segment part F/P. Because segment 2 ends with positioning the rotor in a positive torque position of the starting location of segment one. Part P/P is the movement created by the interaction of two permanent magnets. The segment movement does not need any outside power than that of the flux interacting between the rotor permanent magnet and stator permanent magnet to create the one segment movement of the rotor assembly. At the beginning of the Part P/P segment, the magnetic force causes movement in the rotor to come to a still position at the other end of the segment movement. This physical resting point is where the permanent magnet of the stator assembly to the permanent magnet of the rotor assembly is keeping the movement between the stator and rotor assemblies in the current physical position by the attraction these magnets have toward each other. One could argue that the energy used in segment part F/P was used to load the power offset to create the power free movement of the part P/P movement. This may be true for some of the motor designs, but not all of them as you will see.

Part F/P is the movement between the stator and rotor assemblies as the physical movement is through the second segment of travel. This movement starts at the end of the first segment movement and it moves to the end of the second segment. This is the starting position of another first segment type of movement again for the device. As the flux switch is activated, the flux from the stator assembly interacts with the permanent magnet in the rotor assembly in such a way as to either physically move the rotor assembly to the end of the second segment or it nullifies the repulsion that would normally occur in the second segment of travel depending on the motor design.

A CLOSER LOOK AT FLUX SWITCHING:

There are two major design configurations of rotor assemblies used in this new technology. Each has its advantages and disadvantages.

Power rotor:

This rotor is designed using permanent magnets in every other segment of the rotor assembly. These motors designed with this rotor can create up to twice the power output than the efficiency rotor motor designs. This depends on what efficiency rotor design is used in the motor assembly. When using the Power Rotor, the permanent magnets in the rotor are 50% of the active assembly. In the second segment of travel in the stator assembly, the flux flows through each adjacent set of switches and permanent magnets until it comes around to itself again. This creates a self-contained system of flux flow that will not interact with the rotor magnets as long as the distance between the rotor and stator magnets is greater that the distance between the switching magnets and the ones adjacent to them. The switch device must have the magnetic poles lined up with the permanent magnets to make this work. During this time the rotor acts according to the power of the switching device. If the power is less than the adjacent magnet, then some repulsion will be exhibited on the rotor magnets. If the flux is the same as the adjacent permanent magnets, then the rotor magnets will not have any action on the stator permanent magnets and the rotor will glide through this segment of travel. One configuration with electro-magnet flux switches will put enough electrical energy into it to have about twice the flux as the adjacent magnets in order to create attraction on the rotor magnet to create positive torque on the motor at the same time as supporting the functional ring magnet with the stator magnets. Functionally the rotor see's the stator magnet and the electro-magnet only because the stator permanent magnets have functionally disappeared.

Efficiency rotor:

This rotor design is used for the best efficiency ratings of the motor designs. The rotors are built with permanent magnets in every forth segment of the rotor assembly. This allows for more alternatives that operate the motor with less power per torque output. I prefer the design of the rotor magnets being at about 90 degrees in relationship with the stator magnets. The parallel rotor to stator magnets works the best when they are the same size. Since the magnets react mostly at the poles, when the distance between the poles of the functional magnet is a longer distance from the rotor magnet poles, then the closest poles of the functional stator magnet dominate the closest pole of the rotor magnet. Then the rotor magnet is 90 degrees in relationship to the stator functional magnet, the one side of the functional magnet will push the rotor magnet through the segment of travel while the other pole of the stator magnet will pull the rotor magnet through the segment of travel. Since the pull and push forces are the greatest when the poles of the rotor magnets are closest to the stator magnet poles, then the shorter the functional stator magnet is, the more the force of torque will be to move the rotor. There is one big limit to this because if the functional magnet gets too close to the next functional magnet in the stator assembly, then more flux will flow between the functional magnets instead of interacting with the rotor magnets in the motor. Motor application and design will determine the optimal spacing and magnet lengths in the motor assemblies.

When using the Efficiency Rotor, the configuration where the permanent magnets are in the rotor are 25% of the active assembly, the most common flux switching is done by alternating the flux switching in the motor. The flux switch will cause the flux to flow through the two adjacent magnets to it creating a larger functional magnet that causes the rotation in the rotor assembly. Unlike the power rotor that needs twice the electrical energy to generate twice the flux in order to have full magnetic power to equal the power of the permanent magnets in the rotor, the flux in the functional magnet in this configuration only needs to be equal to one permanent magnet. When the mechanical flux switch is applied to the efficiency design, the mechanical switch becomes part of the functional stator magnet to create forward torque on the motor assembly.

In the power rotor, having the same flux power in the switch as the stator magnets, the flux all moves into the large ring magnet, being self-contained having no torque effect on the rotor assembly. In order to ensure forward torque when using mechanical switching, it is recommended to use motor designs using the efficiency rotors in the design.

Note: When the functional magnet of the Stator permanent magnet is larger than the rotor magnet, then the torque power is less. So, do not look for torque levels as high as what occurred in segment one in these designs.

The Power Rotor configuration creates about twice the power, while this Efficiency Rotor creates a lot more efficiency. The application of the motor determines the best configuration to be used. Each rotor type has its advantages and disadvantages.

NEW: The efficiency rotor can have electromagnets to it in order to generate more power to the motor with less power than the power rotor. That information is in the new chapter added in the front of this book.

WHEN THE FLUX SWITCH IS A PERMANENT MAGNET.

For the power rotor configuration, the easiest way to switch the permanent magnet will be to rotate 180 degrees for each segment of travel. When completing a full segment set of travel, it will have rotated 360 degrees. One way to rotate the switching permanent magnets is a gearing system with one larger gear on the motor shaft. This gear would drive several smaller gears, one for each flux switch, that would stay in sync all the time for the motor operation. The gears would insure the proper rotation on the poles of the switching magnet with the rotor magnets for any possible positive torque on the rotor during the transitional period of the switching action of the flux switch. There will be losses due to repulsion forces that occurs before the full flux switching occurs. Full flux switching is at about 10% of the rotation. This low switching performance may be enough to look for other mechanical approaches to do this.

Note: An extended rotor that overlaps the stator enough to have the gear teeth in the rotor assembly driving the flux switch gear may be an option of doing the switching.

A better way to perform the mechanical switching is to have one larger gear that would drive the flux switch cam assemblies. Each flux switch would have one gear having a piston or cam attached to it. This type of device would be mounted on each gear that would move the mechanical switch faster into and out of the switching position at the same time as to spend more time in the switching position. This would provide more torque for the motor movement.

When mechanical switching is used on Efficiency Rotors that are built with permanent magnets, you need to design it so the flux switching occurs 25% of the active range. Creative ways of switching may need to be designed. Longer motion piston switching may work in order to have the flux switched 25% of the time. This would be the first approach I would use in testing the mechanical switching of this motor concept. NEW: I have since have the dual switching wheel design that has a lot of potential. This is in the proto-type I am currently building in the new chapter I added to the book.

The upside to non-electro-magnet switching is that no electrical circuit is require. The work to rotate the switching permanent magnet is easier than implementing the electronic control circuitry for the electro-magnets. The downside is the additional mechanical structure needed in the motor to rotate the flux switches in sync with the rotor rotation. The mechanical rotation creates a lot of friction and wear on the moving parts with big limitations on speed. So, there may be applications for motors with different flux switching in them.

WHEN THE SWITCHING MAGNET IS AN ELECTRO-MAGNET.

Since the electro-magnet can be switched on and off, then I will turn off the power to it just before the stator magnet would be in repulsion with the rotor magnet. When the rotor is in the position when a positive torque would be created, then I would turn on the power.

For Power Rotors, I would turn up the power on the electro-magnet to output twice the flux of one permanent magnets. In this way I would have enough flux lines in the magnetic force to complete the stator functional ring and have enough flux lines to attract the rotor permanent magnet to travel through its segment of travel.

For Efficiency Rotors, the flux switching occurs every other switch creating larger functional stator magnets creating the torque to move the rotor. The electrical energy to the electro-magnet one needs to provide enough flux lines to match the stator permanent magnets only to do this. This reduces the electrical power usage of the electro-magnet in half. The electro-magnets are permanently mounted into the assembly greatly simplifying the mechanical design, build, and expense in them. The down side is that you need an electrical circuit and power to operate them. To gain high efficiencies requires expensive control circuitry. The advantage to this motor design is that it is capable of much higher speeds and control of the operation of the motor. Safer because when the electrical power is off then the mechanical movement is also off at the same time. If both motors are over-unity, then the electrical flux switching is more desirable to use in most applications.

WHEN SWITCHING IS DONE
WITH CORE MATERIAL.

The core material being moved in and out of the stator permanent magnets to alter the flux stream are similar options as the permanent magnet. When using the rotation method of switching, the speed would be different to rotation of the core material. It would rotate 90% every segment and 180% every segment set to control the flux flow properly. You could use cam/piston action to move the core material in and out of position of the switching. When using core materials rather than magnets for flux switching, you would save the expense of having more magnets in the motor assembly. The downside of core material switching to electronic switching is the additional mechanical structure needed in the motor to slide the core in and out of position of the stator permanent magnet in sync with the rotor rotation. Also, I do not think the core material would provide the overall torque that the permanent magnet switch could provide to the motor assembly. This is because when the switch permanent magnet flux is stronger than the stator magnets, then that additional power would have positive torque on the movement of the rotor assembly.

The following several pages only applies to motors using electro-magnets for the flux switches in the motor device. I will let you know when it applies to other switching types.

When the flux switches are energized using the new technology, the interactions between the permanent magnets in the rotor and the stator are changed. There are two different ways these changes are made depending on the mechanical makeup of the motor. The first is change by sizing and the second is change by dual function. Another thing to keep in mind is that the electro-magnets need to have cores that have little to no interaction with the permanent magnets when the power is off. For now, the core material will be air. Since I need to wrap the magnetic wire used to build the electro-magnets, I will use copper tubing, aluminum rods, plastic, or other types of materials.

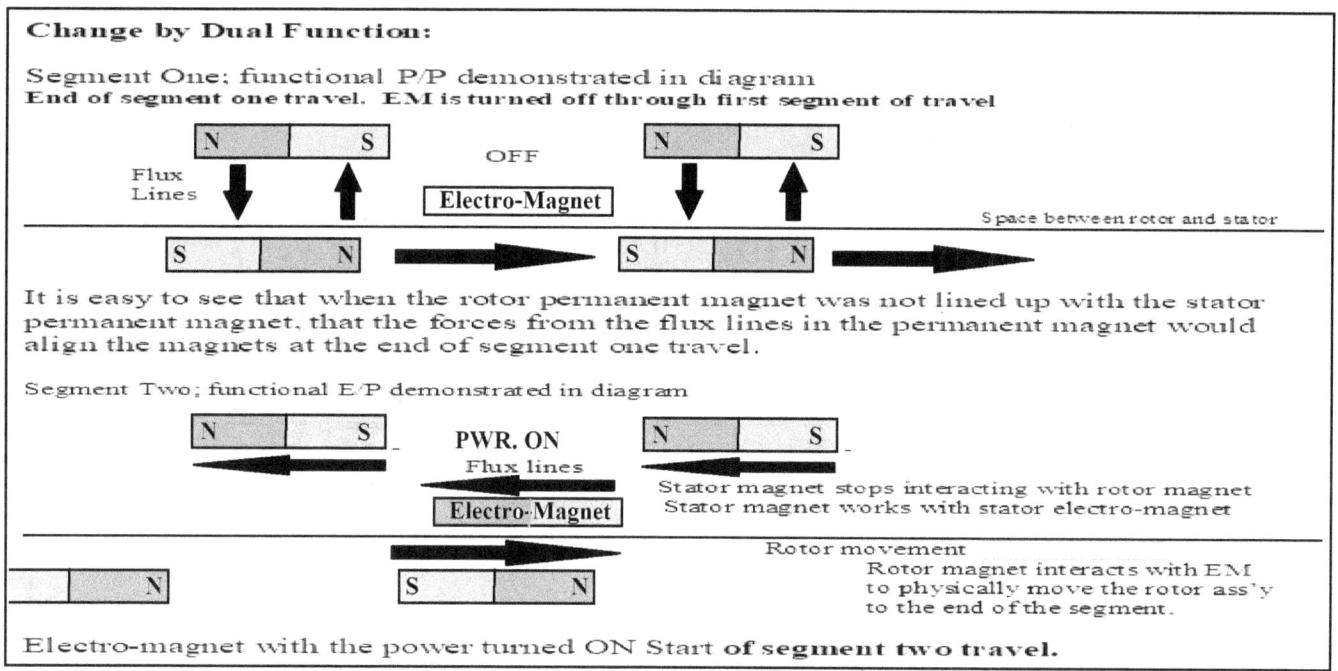

EM is turned off through first segment of travel. EM is invisible to PMs

This is easy to see that when the rotor permanent magnet is not lined up with the stator permanent magnet, that the forces from the flux lines in the permanent magnet would align the magnets at the end of segment one travel.

Segment Two; functional magnet is demonstrated in diagram; the two PMs and the EM create one large electro-magnet with the power turned ON.

The dual function happens in this segment with the EM doing two different things at the same time. The function of the individual permanent magnets between the rotor and the stator assembly is broken when the power of the electro-magnet is turned on. So, this releases the force that was keeping the permanent magnets aligned to each other at the end of the first segment movement. The electro-magnet creates new connections with the flux lines that are created to start the physical movement through segment two until it comes to the end of the movement at the end of the second segment of travel.

Note: When the power to the E.M. is turned on it completely changes the dynamics of the assembly causing all the permanent magnets to work with it in all new positive ways to maximize the movement of the device down the track or around the circle.

Note: When the power to the E.M. is turned off the back EMF that is created causes the EM to have the opposite polarity. The Permanent Magnet in the device is now in a location where this back EMF causes a positive torque to move the device down the track or in a circle. It is a free push.

Below several Parallel PFPMMDs are lined up linearly to create one side of a track. The other side of the track looks similar.

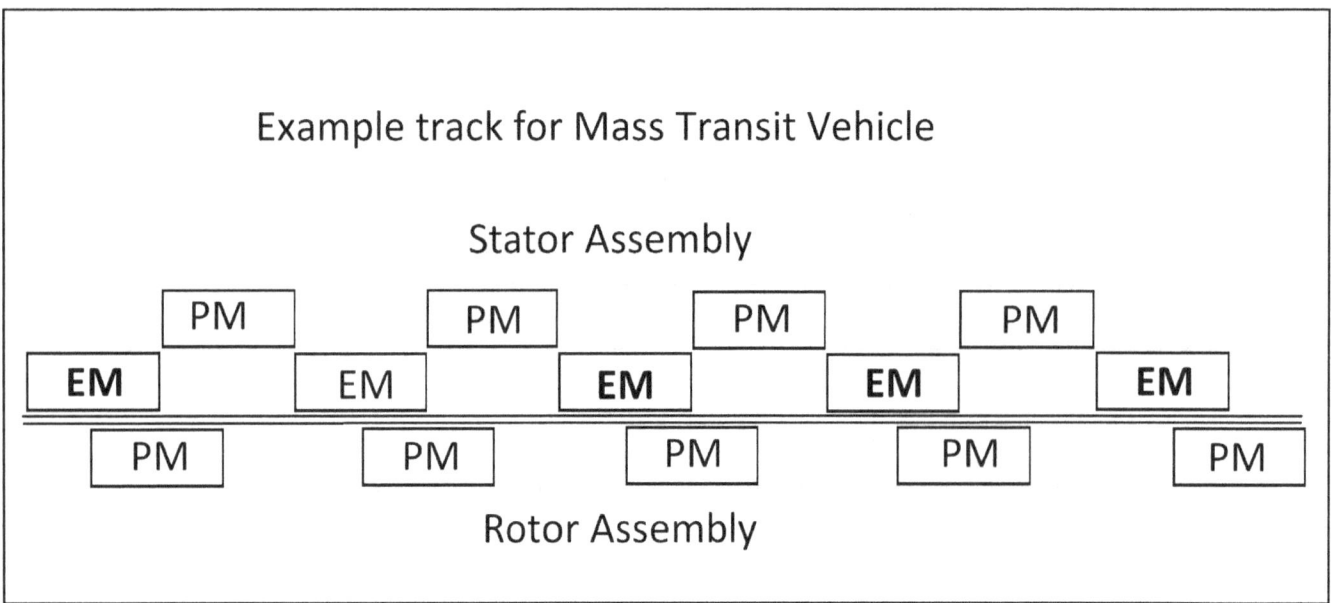

This interaction is so unique that it needs reinforcing by describing it again. The energized electromagnet during the functional F/P movement greatly reduces or disables the interaction of the individual permanent magnets in the stator to the individual permanent magnet in the rotor assembly. It does this by creating two new interactions.

The first interaction is between the electro-magnet in the stator with the permanent magnets in the stator assembly. The flux going through the flux switch causes the flux of the adjacent permanent magnets to flow through it creating one large functional ring magnet in this layer of the technology. This one large functional ring magnet does not interact with the rotor permanent magnets.

The second new interaction is between the electro-magnet in the stator to the permanent magnet in the rotor assembly to that of attraction. This attraction moves the rotor through segment two. For this to work, the power that the electro-magnet creates, needs to produce more flux lines than what is needed to put the permanent magnet layer in the stator to sleep in respect to the rotor. So, the additional flux lines generated in the electro-magnet create an attraction for the permanent magnet in the stator assembly to create one segment of movement of travel. At the end of the physical movement created by the mechanical components in the functional F/P assembly positions the physical placement of the assemble in the correct physical position to start the movement of the functional P/P all over again.

This completes a segment set movement. One thing to note is that ½ of the segment set's movement is made up of a permanent magnet to permanent magnet interaction. The other ½ is by the electro-magnet to permanent magnet interaction. This in itself will improve the efficiency of the new motor technology over the conventional electric motor that uses all electro-magnets in its design.

Together part P/P and part F/P produce a very efficient electro-mechanical device producing mechanical movement. The fact that, in the dual function designs, three permanent magnets to one electro-magnet which only uses electricity either 25% or 50% of the travel for the device means that the motor has a high efficiency rating. When you couple this with my new resonant power circuitry for the motor, the motor will have some of the best performances ever achieved for electric motors.

Overlapping the two stator layers:

With the permanent magnets in layer 3 over lapping electro-magnets in layer two of the three-layer technology, the power output can be improved. Because of the geometry of the electro-magnets in the stator to that of the permanent magnets in the stator assembly, it may not take that much power in the electro-magnets to attract the flux from the permanent magnets to start flowing through each other to create that one large functional ring magnet in the stator. Once the flux lines go into this mode, the permanent magnets in the rotor will have little affect from the permanent magnets in the stator assemblies. If the flux of the electro-magnets match that of the permanent magnets in the stator assembly, then the torque would be nothing during segment two, the rotor would glide through this segment. If you choose to operate the flux switch at the equal flux level as the stator magnet, you will either need other disks in the motor to offset this gliding point in order for the full motor assembly to have torque during the full 360 degrees of rotation. If you are using a disc for the rotor assembly, then you could install another stator on the other side of the rotor. This stator would offset the segments so that the motor package would always have a segment one in operation to provide the torque through 360 degrees of rotation. A percentage of overlapping can create a more desirable design configuration for the application. The best design configuration will most likely be between zero to full layer offsets. The application will determine how to optimize most of the motor designs.

The design goal for overlapping should be for improving the motor's performance for its application. Having the magnets from the stator to be close to the rotor permanent magnets is to maximize the force used in moving the rotor through segment one without sacrificing the movement of the rotor in the second segment. In the second segment movement, the electro-magnets need to greatly reduce the segment 1 type of interaction between the permanent magnets in the stator with the permanent magnets in the rotor when the power to the electro-magnets is turned on. So, in most designs, there needs to be some distance between the stator and rotor permanent magnets. The electro-magnets need to be as close to the rotor permanent magnets as possible to maximize the movement of the rotor in the second segment of motor movement. The optimum distance will need to be evaluated for each motor type. The overlapping approach supports and compliments the three-layer technology for designing motors.

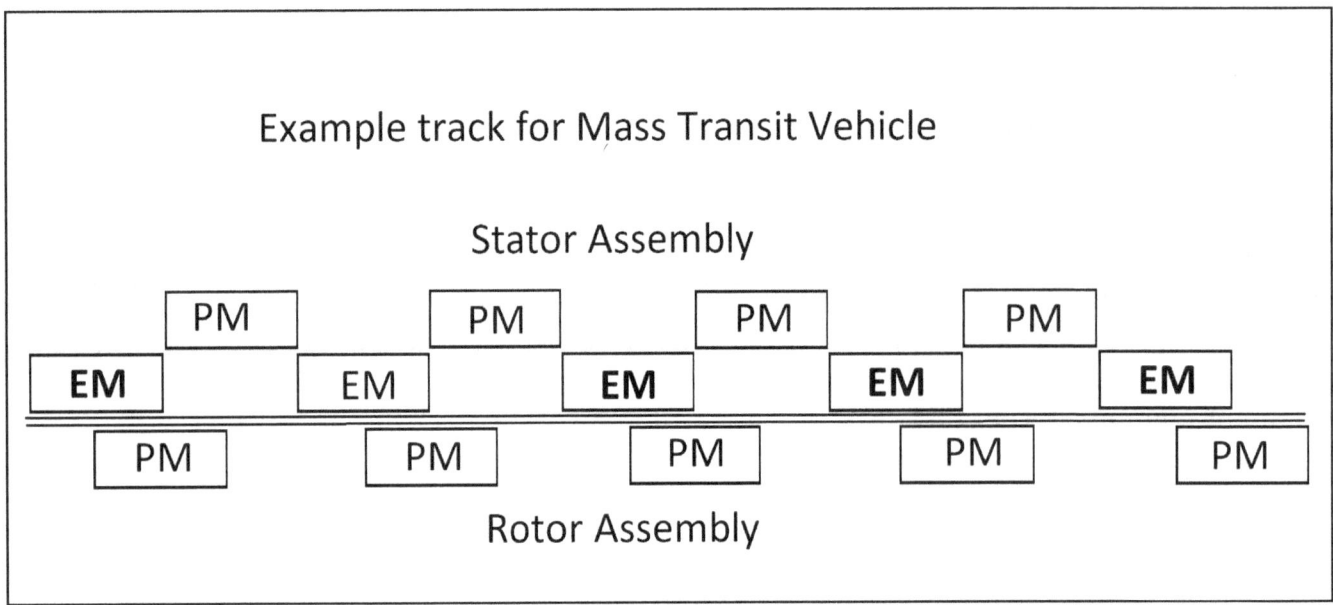

Example track for Mass Transit Vehicle

Stator Assembly

Rotor Assembly

Change by Sizing: See following figures

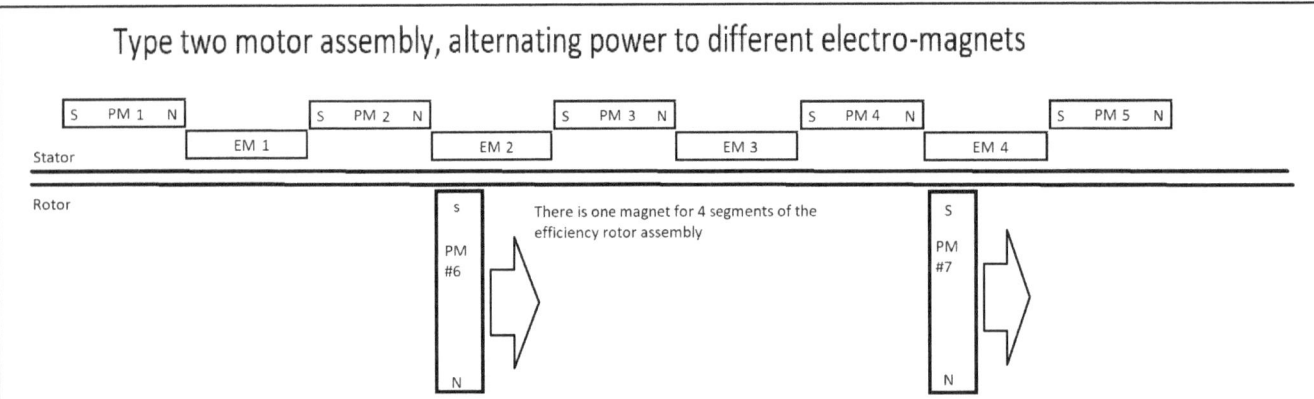

Type two motor assembly, alternating power to different electro-magnets

There is one magnet for 4 segments of the efficiency rotor assembly

The electro-magnets EM1-EM4 are all turned off. The south pole of PM6 is pulled to line up with the north pole of PM2. The south pole of PM 7 is pulled to line up with the north pole of PM4.

Rotor moved to next segment

Type two motor assembly, alternating power to different electro-magnets

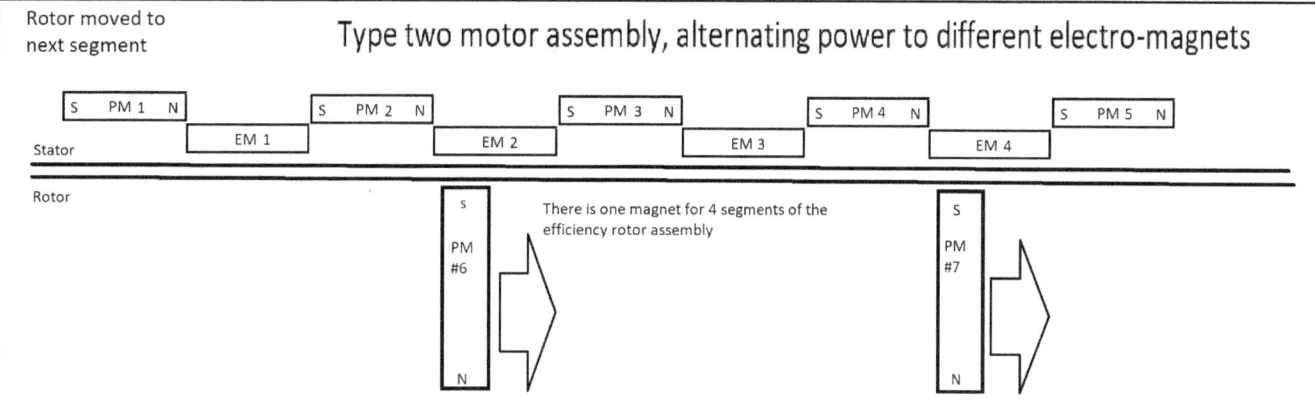

The electro-magnets EM2 and EM4 are turned on and EM1 and EM3 are turned off. The south pole of PM6 is pulled to line up with the north pole of EM2 and the south pole of PM7 is pulled to line up with the north pole of EM4

Rotor moved to next segment

Type two motor assembly, alternating power to different electro-magnets

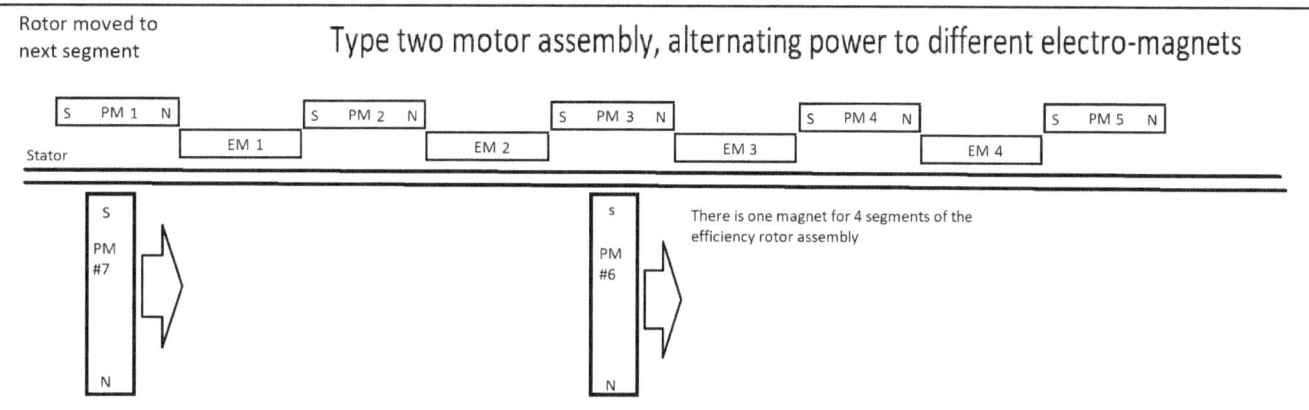

All the electro-magnets are turned off. The south pole of PM6 is pulled to line up with the north pole of PM3 and the south pole of PM7 is pulled to line up with the north pole of PM1.

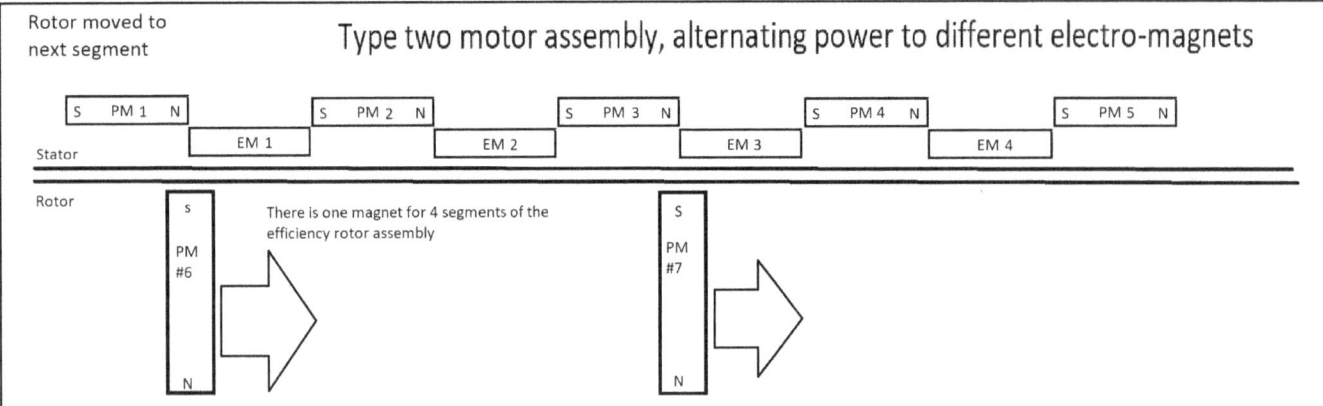

Type two motor assembly, alternating power to different electro-magnets

Rotor moved to next segment

The electro-magnets EM1 and EM3 are turned on and EM2 and EM4 are turned off. The south pole of PM 6 is pulled to line up with the north pole of EM3 and the south pole of PM7 is pulled to line up with the north pole of EM1.

The design that uses this "change of sizing" would have ½ of the permanent magnets in the rotor than that of the stator assembly. Reducing the permanent magnets in half for the rotor assembly would mean that there would be one permanent magnet in every 4th segment of the rotor assembly. The function of the rotor permanent magnet stays the same but the stator magnets change their functional size. The way this happens is that the electro-magnet creates an electrical field that pulls the flux lines from the magnet in front of it through itself and then through the magnet behind it. This creates a functional magnet consisting of two permanent magnets with one electro-magnet between them. This functional magnet is about three times longer than when the electro-magnet is turned off. So now the rotor permanent magnet sees the north pole at the end of one permanent magnet and the south pole at the other end of the other permanent magnet. This will cause the rotor to start moving to a new resting place. The end of the electro-magnet just so happens to be at the end of the second segment. This is where to power is turned off to the electro-magnet. This type of operation requires that the alternating electro-magnets to be turned off. If they were not turned off, then the permanent and electro-magnets in the stator assembly would create one big ring magnet that would not interact with the rotor permanent magnet through this segment of travel. This change by size configuration would only be able to achieve 50% to 75% of the power of the Power Rotor motor types that use all the electro-magnets at the same time.

The power cycle for each individual electro-magnet is ¼ of the time. So, one half of the electro-magnets are turned on during the second segment of travel while the other half are turned off. These electro-magnets that were turned off during the second segment are turned on during the fourth segment of travel keeping the other electro-magnets power off during the motor's operation. Also, the power used in the electro-magnet is ½ of that of the Power Rotor design. This means the electrical energy is cut to 25% when using this configuration to operate the motor. The efficiencies made by having five permanent magnet interactions per one electro-magnet action operating in this mode of the "change by size" motor design would be great.

Note: In some rare applications or conditions, the electro-magnets that are normally off, may require a current in the opposite direction to repel the flux lines from the permanent magnets from flowing through them creating one large ring magnet. The back EMF that is generated with electro-magnets creates this same polarity we want to keep the ring magnet from being generated during the motor operation. If the back EMF was not enough, then the reverse power would be needed. With proper motor design of magnet placement, this should not be an issue. I wanted to bring this up so that it is not over-looked through motor design efforts. Remember that we want the functional larger magnet configuration when using this Efficiency Rotor design.

These two different types of PFPMMDs becomes the building blocks for many different electro-mechanical devices. Adding PFPMMDs to the motor assembly is a simple thing to do to keep the movement continuing as far as you want it to. Many devices can be created from this simple building block. If the motor design uses a circular design like that of disc assemblies, these assemblies would be built into function sets of discs. The functional set would contain the three layers of the new technology built into them. The motor could be built with multiple sets of functional disc assemblies with each rotor having two stators. The rotor magnets may need to produce more flux in them compared to the stator magnets. The placement of these assemblies could be offsetting so that their peak torque points in relationship to each other are staggered. In this way the total motor assembly would have more of a constant torque on it during the full 360 degrees of rotation. This is more important in the motor start up activities or in slow moving rotation applications of the motor.

DESIGNING WITH THIS NEW THREE-LAYER TECHNOLOGY

Design the motor into the application reducing the overall number of parts used in the device.

The motor does not always have to turn a shaft. Movement in the outer circumference of the motor provides higher torque for many applications.

When using mechanical switching devices to power the motor, use those devices for a switching signal and not the power of the switching. The switching signal should drive another device like a solid-state relay. This would reduce the wear on the switching mechanical hardware that would be used in the device.

If you cannot use wireless switching, then consider optical switching in place of conventional mechanical switching. When you use optical switching, use trigger points for the switching as far out on the circumference of the motor as possible in order to supply the best clear switching points for the motor control circuitry as possible.

Use computer speed electronic control components to be able to make adjustments to the motor timing as fast as possible.

Use low resistance wiring to improve the efficiency of the motor.

Optimize the electromagnets for core materials that optimizes both the P/P and F/P functional operations of the motor assembly. Look for materials to do this that are better than air core for these motor designs.

Build an adjustable testing motor assembly that can be easily modified and tested in order to optimize the motor design.

A better way in designing motors that use as little power as possible. That would be to only use the electro-magnet power when it is needed to move the device. In this type of circuit, the power to the electrical circuit during segment two would have a duty cycle attached to the on time. This circuit would very the speed of the device as the duty cycle changed. For a two-segment configuration, we are looking at full speed when the duty cycle for the segment set operating at 50%. Remember that this is because 50% of the movement does not require power to move the assembly in the first place. When the device P.M. lines up with the E.M. the power is turned off. The four-segment configuration would have a duty cycle of about half that of the two-segment configuration.

Whenever possible, use the Resonant oscillating power control circuit for the motors control. There are many options in creating a variable capacitor circuit. Switching in different size capacitors into the circuit can be done using switching chips. If a combination of voltage and capacitance values are used in the resonant oscillating circuit, then the maximum voltage ratings of each component need to meet the range used in the circuit.

RESONANT OSCILLATING
TANK POWER CIRCUITRY:

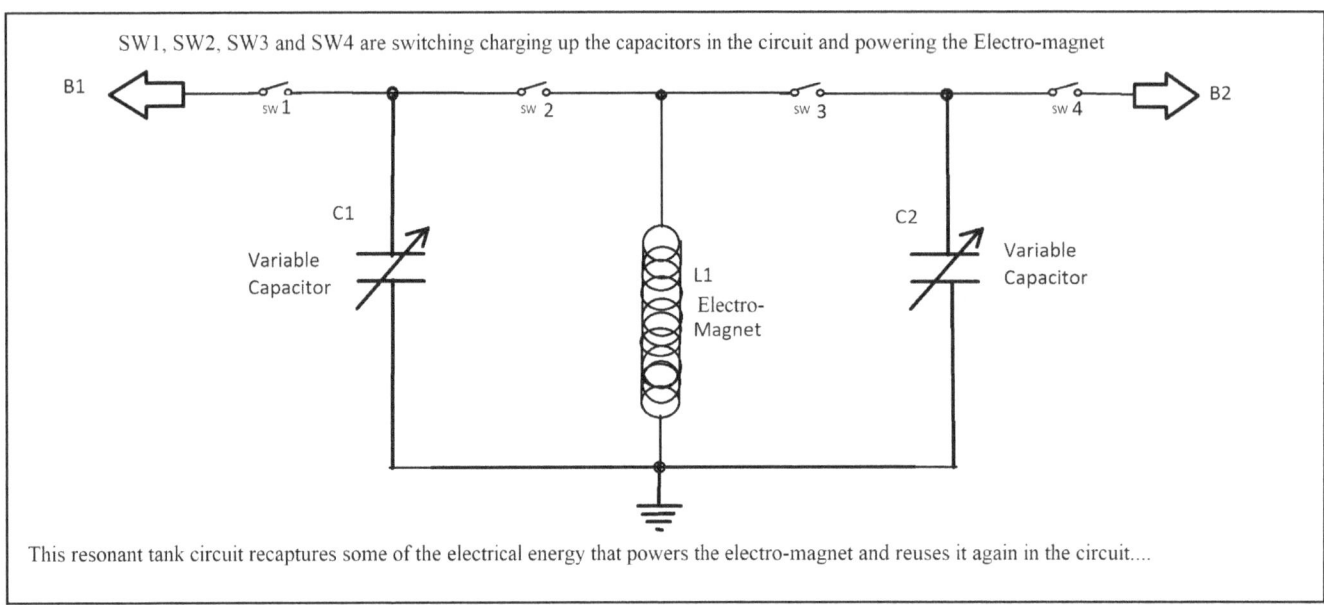

SW1, SW2, SW3 and SW4 are switching charging up the capacitors in the circuit and powering the Electro-magnet

This resonant tank circuit recaptures some of the electrical energy that powers the electro-magnet and reuses it again in the circuit....

S1, S2, S3 and S4 are switches for this circuit

For most applications B1 and B2 could be the same power source.

> S1 and S3 are closed in segments 1 and 2 while S2 and S4 are open
>
> S2 and S4 are closed in segments 3 and 4 while S1 and S3 are open
>
> The tank circuit operates like a pendulum where the electrical energy swings back and

forth between a coil and a capacitor.

The tank power supply circuitry that I am describing here was first written in another paper I wrote in 1986. This design would be the best efficient circuit to operate the new motor device with. I will summarize the advantages of this circuit. When you have a coil and a capacitor tied together, they will have a resonant frequency where the current flows back and forth between them. These circuits are found in all the old tube TV sets and many other circuits. The current flow is very efficient moving in this type of a circuit. The circuit has only wire resistive and some induction losses in it. When the current flows back and forth in the coil, a magnetic field is created. The polarity of the magnetic field changes direction as the current flow changes direction in it. The electro-magnets would replace the coil in the circuit. The back EMF occurs when the coil is releasing its energy back into the capacitor. This back EMF is in the opposite polarity as when the power is going into the coil. When the back EMF starts, the physical position of the rotor has moved into a location that needs this reversed polarity to create positive torque on the motor assembly. The capacitor then captures the energy minus the resistive and induction losses back into the capacitor. The switching is made to then charge the capacitor back to its full potential by connecting it to a power supply. The reason I have two variable capacitors instead of one is so that the capacitors can be topped off to a full charge before they are used again. Note that when switches to the capacitor is open, the capacitor locks in the power it captured and holds it for as long as you want to hold it. That is why I can have two capacitors in the tank circuit and it will operate at some percentage of efficiently of that of the TV tank circuit. So, for each operating condition of the motor, the capacitor value is adjusted to the resonant point. This point is the most efficient energy usage point of the motor assembly.

A control circuitry needs to be used to control all the variables of a changing resonant point of the motor assembly over a wide range of operating conditions. Together the motor high efficiencies along with the power and control circuitry would result in some of the markets highest efficient motors.

Even though the Resonant oscillating power control circuit was not designed for the three layered electro-mechanical devices, with a few modifications, the power supply circuit should be the best efficient circuit to operate the mechanical devices with this technology. This circuit could be easily modified to operate other existing electric motors on the market today. With measuring the RPMs, rotor location and current draw of the motor, the circuitry could do a great job of improving the efficiency of the motor's operation.

The resonant point I refer to in my paper is the point where the current going into the electro-magnet stops flowing is at the point of one segment of travel. At this point the motor will be operating at one speed using one supply voltage and one capacitance value when this happens. If you want the motor speed to be at a different value, then you will need to change something else in the circuit. The best thing to change is the voltage the control circuit is operating at. The general case is, the higher the voltage, the higher the speed the motor will be when it reaches the resonant point.

The Flow Through motor and the resonant power circuits were designed by me several years earlier. The earlier writings show many design enhancements and design options for different applications that can also be utilized for this current motor design as well. The flow through motor was never patented. Neither was any of my motor control circuitries. So, I would have no problems using ideas from my earlier motor designs today. The new motor technology has more technological differences in it than the flow-through motor design. The improved technology makes the new three-layered motor design technology more efficient for a multitude of configurations. Even the modified tank control circuit is unique to other tank circuits in the fact that it is used in the movement of a mechanical device rather than used in a stationary resonant assembly which in the past was used for electronic circuit applications.

The power rotor circuit could operate with one resonant circuit to control all the electro-magnets at the same time. The efficiency rotor would require a minimum of two resonant control circuits since the timing is different for some of the electro-magnets in the motor for this assembly. The maximum control would be one resonant control circuit for each electro-magnet. I do not see the need for that kind of control if the electro-magnets are designed to have equal resistances and inductance.

The following drawing as another new option that actually simplifies the TANK circuit for operating electromagnets in many motor configurations.

This circuit needs to have a logic circuit to function the seven switches. One mechanical arm can be used to signal the size of the segment. Four pickup devices can be spaced with the mechanical arm in order to have the four segment signals. These four signals will be sent into a simple logic sircuit in order to have a drive out seven switch signals. These signals will drive solid state relays to provide the power requirements of the electromagnets.

There could also be a manual switch that would switch between this circuit shown and a dirrect power circuit that could be used for the start-up sequence of the motor in the first proof-of-concept motor. The start up could be incorporated in a more complex power/control module of a production type of motor.

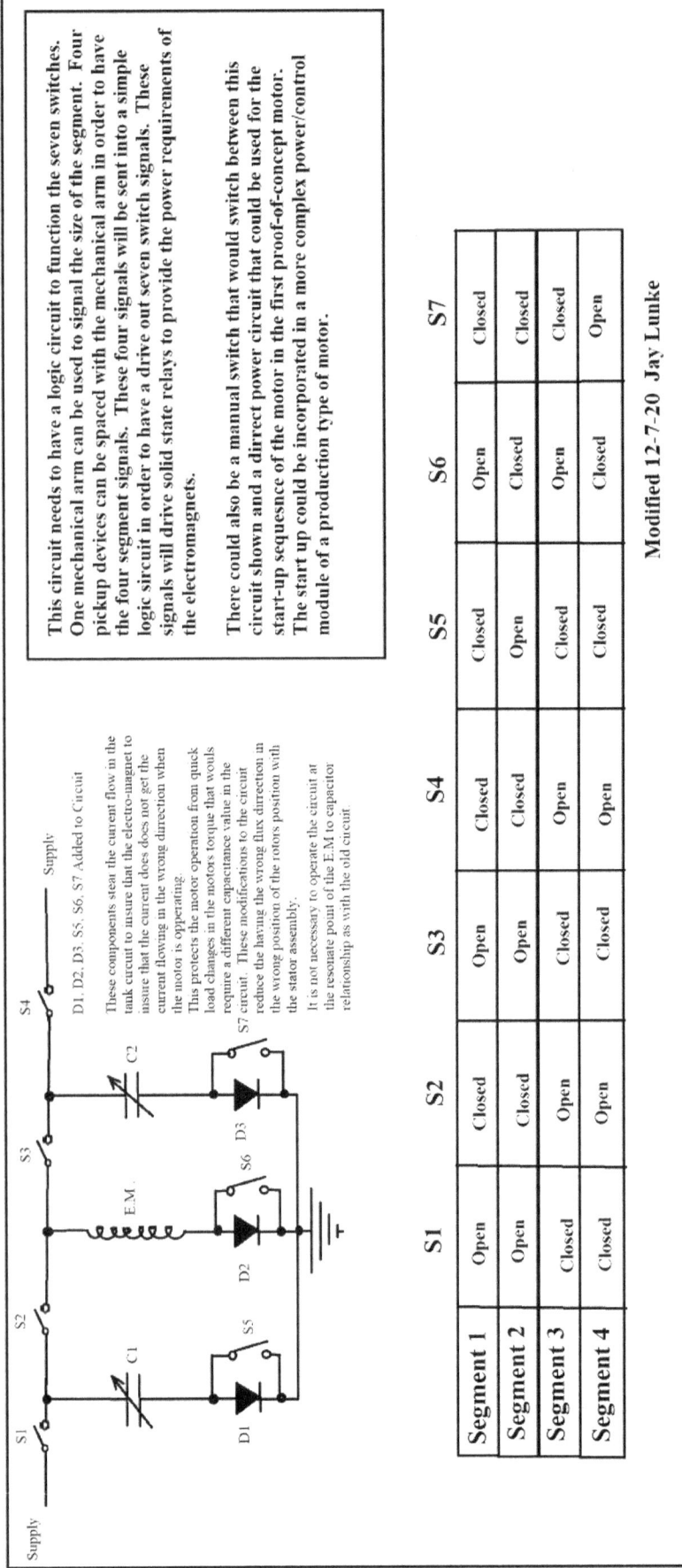

D1, D2, D3, S5, S6, S7 Added to Circuit

These components stan the current flow in the tank circuit to insure that the electro-magnet to insure that the current does does not get the current flowing in the wrong direction when the motor is opperating.

This protects the motor operation from quick load changes in the motors torque that wouls require a different capacitance value in the S7 circuit. These modifications to the circuit reduce the having the wrong flux dirrection in the wrong position of the rotors position with the stator assembly.

It is not necessary to operate the circuit at the resonate point of the E.M to capacitor relationship as with the old circuit

	S1	S2	S3	S4	S5	S6	S7
Segment 1	Open	Closed	Open	Closed	Closed	Open	Closed
Segment 2	Open	Closed	Open	Closed	Open	Closed	Closed
Segment 3	Closed	Open	Closed	Open	Closed	Open	Closed
Segment 4	Closed	Open	Closed	Open	Closed	Closed	Open

Modified 12-7-20 Jay Lunke

Alternative power circuit to explore: Reduced-back EMF circuit.

The way this power circuit works, is by knowing how long it takes for the electro-magnet to release its energy through what is a normal back EMF event. In would further improve the circuit. The way I would do this would be to float the electro-magnet off of ground. I would either operate the motor with a negative power supply or I would leave the positive side of the power always connected to the supply and switch the other side of the electro-magnet to the ground potential. What this would do is to reduce the reverse EMF. A resistor could be added in the circuit to reduce the current flow in the coil. When the electro-magnet is first connected, the electrical current flows from the negative side through the electro-magnet to create a magnetic field with the poles fixed. When the ground side is opened up, the current is still flowing in the same direction and the voltage and current becomes smaller and smaller until it stops altogether. The ideal would be that the magnetic field reduces size but it remains in the same direction. This has not yet been proved out. Even if the back EMF is reduced, this would be an improvement for the motor's performance. The resistor would slow down the collapsing of the magnetic field. The timing of the switching on and off of the electro-magnet becomes the critical part of the motor operation in order to optimize the motors performance. The amount of electronic circuitry will be less than the resonant circuit. The power supply voltage level and/or the switching on time can be used to adjust the speed of the motor.

Simplest Three Layer Device:

The following motor is the simplest three-layer motor device using the new technology. It only uses three magnets in the motor. The disadvantage of this motor over the other three-layer motors is that there are no other stator permanent magnets for the electro-magnet to influence in to one large magnet that does not care about the rotor permanent magnets anymore. Instead, the electro-magnet blocks the interaction of the stator permanent magnet with the rotor permanent magnets by directing the flux of the stator magnet through the electro-magnet. This reduces the efficiency of the design. This motor does show the other advantages of the three-layer technology. It is a lot simpler to build this motor than the other configurations. The simpler designs will also always have a place in smaller applications. The motor does have ½ of the rotation using only the two permanent magnets making the movement. The other 50% needs to have the electro-magnet to become a magnet that will block the stator permanent magnet from interacting with the rotor permanent magnet. The electro-magnet in this configuration would need twice the power to create twice the flux in it so that 100% of the flux from the stator magnet goes through the electro-magnet and 100% of the flux from the rotor magnet goes through the electro-magnet. Through the segment set of rotation, the functional P/P and F/P movements are built up by three permanent magnets and one electro-magnet. This is because in this segment set movement, the rotor permanent magnet is used twice.

SIMPLEST PARRELLEL MOTOR DEVICE

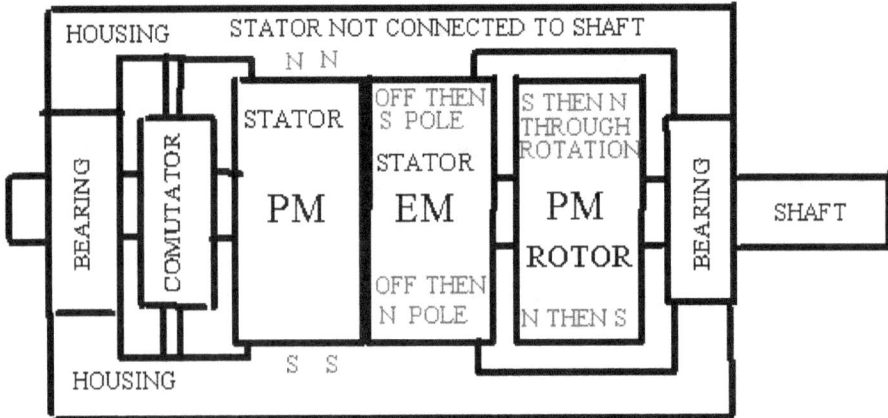

NOT DRAWN TO SCALE; DRAWN TO SHOW THIERY OF OPERATION OF

SIMPLEST PARRELLEL MOTOR DEVICE USING TWO PERMANENT MAGNETS AND ONE ELECTRO-MAGNET
WHEN THE ELECTRO-MAGNET IS TURNED OFF THE SOUTH POLE OF THE ROTOR IS ATTRACTED TO THE
NORTH POLE OF THE STATOR PERMANENT MAGNET CAUSING 180 DEGREE ROTATION OF THE ROTOR.
WHEN THE ELECTRO-MAGNET IS TURNED ON IT PROVIDES A RETURN PATH FOR THE STATOR PERMANENT
MAGNET AT THE SAME TIME PROVIDING THE OPOSIT POLE FOR THE ROTOR PM CAUSING 180 DEGREE
ROTATION OF THE ROTOR ASSEMBLY THE ELECTRO-MAGNET RUNS AT 50% DUTY CYCLE WHILE THE
ROTOR HAS POSATIVE TORQUE ON IT FOR ROTATION FOR THE FULL 360 DEGREES OF ROTATION.

NOTE: THE COMUTATOR IS ONLY ONE OF MANY OPTIONAL DESIGNS TO
TURN ON AND OFF THE ELECTRO-MAGNETS.

JAY.   .LUNKE
DRAWN 1-9-2007

Drawing shows the 3-layer technology using disc. Assemblies. The disc. Assemblies are side by side interacting with each other rather than having different layers of the technology operating at different circumferences in reference to the center point of the motor assembly. The disc. assembly can use more standardized rectangular shapes for the permanent magnets and still have the close proximity to the other disc assemblies. This means a reduction of manufacturing costs for the devices. The maintenance of the disc assemblies is easier to perform than most of the other motor designs.

The three-layered technology can become even more efficient when less repulsion forces are at work in the device and more attraction forces are at work in them. This is done by reducing the active electro-magnets by 50%. The electromagnets alternate being active with the other electro-magnets by 50%. So, each individual electro-magnet has a 25% duty cycle. The torque of the motor is reduced, but the energy to operate the motor is reduced even more because of the efficient way it operates. This technology works for almost all of the three-layer motor design configurations.

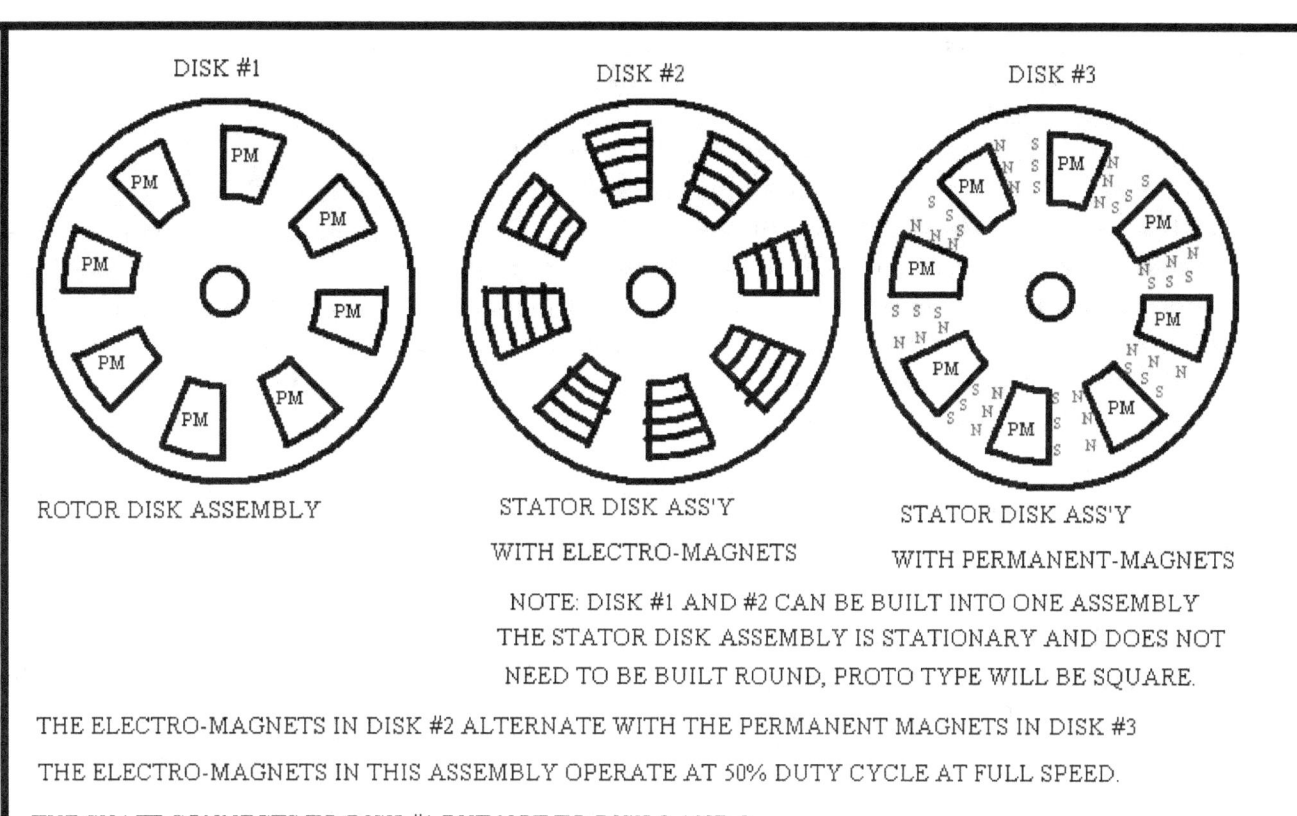

DISK #1

DISK #2

DISK #3

ROTOR DISK ASSEMBLY

STATOR DISK ASS'Y

WITH ELECTRO-MAGNETS

STATOR DISK ASS'Y

WITH PERMANENT-MAGNETS

NOTE: DISK #1 AND #2 CAN BE BUILT INTO ONE ASSEMBLY
THE STATOR DISK ASSEMBLY IS STATIONARY AND DOES NOT
NEED TO BE BUILT ROUND, PROTO TYPE WILL BE SQUARE.

THE ELECTRO-MAGNETS IN DISK #2 ALTERNATE WITH THE PERMANENT MAGNETS IN DISK #3

THE ELECTRO-MAGNETS IN THIS ASSEMBLY OPERATE AT 50% DUTY CYCLE AT FULL SPEED.

THE SHAFT CONNECTS TO DISK #1 BUT NOT TO DISK 2 AND 3.

50% of the time the power is off and the torque is provided by the interation between disk 1 and disk 3 magnets.

50% of the time the electro-magnets are on and torquing the rotor assembly.

The resulting motor is low cost and high efficiency.   Jay Lunke drawn 1-4-2007

Disk PFPMMDA Applications:

These motor assemblies can then be used in several applications. One of the nice features of the disk assemblies is that when the electro-magnets are turned on to a power level to equal the stator disc magnets, the permanent magnet flux will travel from permanent magnet then through an electro-magnet then to the next permanent magnet in the disk assembly and so on until the flux returns to the magnet it started from. This easy route for the stator permanent magnets to travel greatly reduces or eliminates the stators interaction with rotor permanent magnet assemblies. Since the electro-magnets are on the stator assembly with the stator permanent magnets, any forces of attraction or repulsion between them should not affect the movement of the rotor assembly. The electro-magnets can produce more flux creating attraction to the rotor magnets during the segment two rotation. The discs can have the rotor permanent magnets in every forth position which would have all of the Efficiency Rotor design options we have already discussed for it.

I will compare the performance between the two different types of disk motor design types here. Type one is a Power Rotor or power disk assembly. This is when the permanent magnets in the stator are the same number in the rotor assembly. Type two is the Efficient Rotor disk assembly. Type two has half the permanent magnets on the rotor assembly as those found on the stator assembly.

Note also that layers one and two could even be over-lapping each other at some percentage to enhance the performance of the motor assembly. My best guess is a 50% overlap for Power Rotor designs since the electro-magnet has dual functions when energized. It will be closer to 100% overlap for the Efficiency Rotor motor designs. The closer the stator permanent magnets are to the rotor magnets, the greater the torque is from the motor.

Since my main goal is creating a system of maximizing the usage of permanent magnets and reducing the usage of electrical energy used in the electro-magnets to produce the most energy efficient motor assembly, then I favor the Efficiency Rotor motor configuration even though it would need to be a larger motor assembly for the same power output of the Power Rotor motor assembly.

In the Efficiency Rotor motor, there are two stator permanent magnets for every one permanent magnet in the rotor assembly. The electro-magnets have a 25 percent duty cycle compared to the 50 percent duty cycle of the Power Rotor motor design. The electro-magnets will run cooler in the Efficient Rotor motor using 25% of electrical power compared to the Power Rotor design. The power will be somewhere between 50% to 75% of the power of the Power Rotor motor assembly.

Let's look at the condition when the permanent magnet of the rotor assembly is lined up with the permanent magnet on the stator assembly. When the power is first turned on the electro-magnets, the rotor magnets have equal pull between the two adjacent magnets. Since the Power Rotor takes more time to be brought up to its full flux level, the Efficiency Rotor electromagnets only needing half of the power to reach the full flux potential thus reaching that level faster. This difference has a greater effect on the motors the faster they rotate.

With the Efficiency Rotor motor assemblies, when the power is first turned on to every other electro-magnet, the permanent magnet in the rotor assembly is attracted to the closest powered electro-magnet. Since the magnet on one side of the permanent magnet is turned on while the electro-magnet is the same distance from it is turned off the Rotor permanent magnet has only the attraction of the electro-magnets that are turned on and moves toward that one.

Testing of both Power Rotor motors and Efficiency Rotor disk motor assemblies are needed. People should design proto-types common stator assemblies so that different rotors can easily be built into each type for faster testing purposes.

There should be applications for both motor types. The Efficiency Rotor motors will be used for stationary systems. Since the Power Rotor motor has about twice the power, it would be used in motor applications needing more power like automobiles. If you needed even greater power, you could use one of the motor assemblies I created in the flow-through motor design several years ago.

When the test results show that the technology is sound, then it is time to do serious work to build several mechanical devices and start testing them in laboratories to get an accurate measurement of their efficiencies.

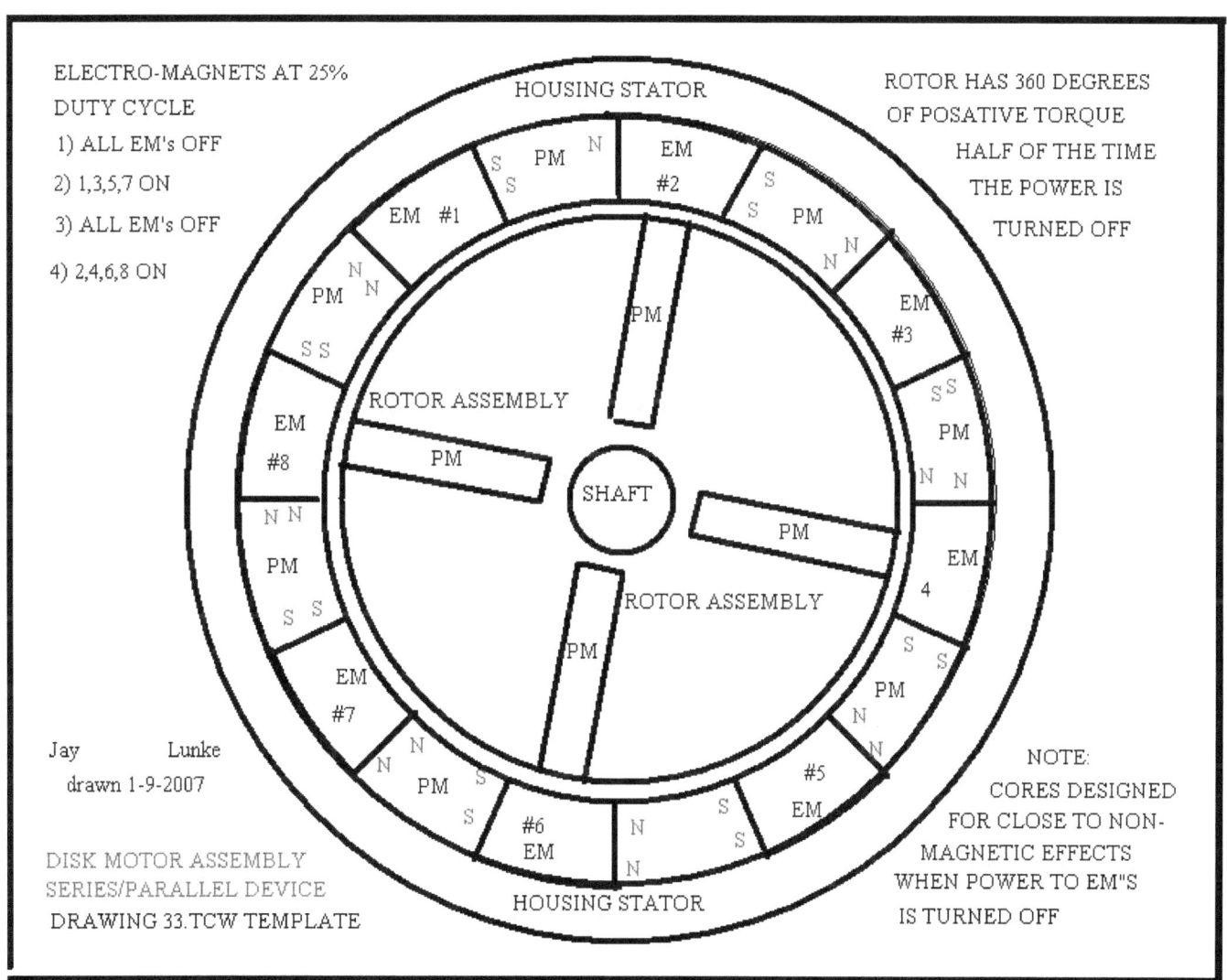

This three-layered motor technology shows maximum over-lapping of two of the layers. The technology of changing the behavior of the two different layers is still active if the power in the electromagnets is greater than that of the permanent magnets. The power to the first set of electro-magnets is switched on in segment two and are turned off in segments 1, 3 and 4. The second set of electro-magnets that alternate with the first set of electro-magnets are turned on in segment 4 and turned off in segments 1,2 and 3. The electro-magnets that are turned on create a larger functional magnet interacting between the stator and rotor assemblies. This does change the interactions in the three-layered technology from Power Rotor to Efficiency Rotor functional configuration already discussed in this paper. Instead of the permanent magnets ignoring the rotor magnets when the electro-magnets are turned on, it changes the size of the permanent stator magnet assembly in order to create the desired forward movement by the rotor again. The orientation of the magnets is different in the rotor in order to show you the versatility of this three-layer technology. So, in one segment the movement is permanent magnet to permanent magnet interaction. In the next segment it is permanent magnet to magnetic assembly created by two permanent magnets and one electro-magnet. This functional magnet made up of two permanent and one electro-magnets is a hybrid magnet assembly that only becomes functional when the power is turned on to the electro-magnet. So, a segment set movement would be made up of five permanent magnets and one electro-magnet. This is the first motor that needs to be tested because it has the most promise of all of them. A circular rather than linear motor would be easier to test as well.

The motor designs that have the electro-magnets operating in the attraction mode with the permanent magnets are the best ones to preserve the strength of the permanent magnets. The one big nock on using permanent magnets in a motor is because the electro-magnet fields operating in opposition to the flux lines of the permanent magnets. This decreases the magnetism of the permanent magnet over time. The attraction mode does not do that. With that in mind, my motors can be used in the larger applications where the old-style permanent magnet motors have not been used. Also, the rare earth magnets that are built today are more robust than the permanent magnets that were created with other materials years ago.

Here is just one of the many mechanical devices that this new technology can be used in. I like the idea of incorporating the motor components of this new technology into the application device itself. This approach to designing has so many advantages it is hard to list them all.

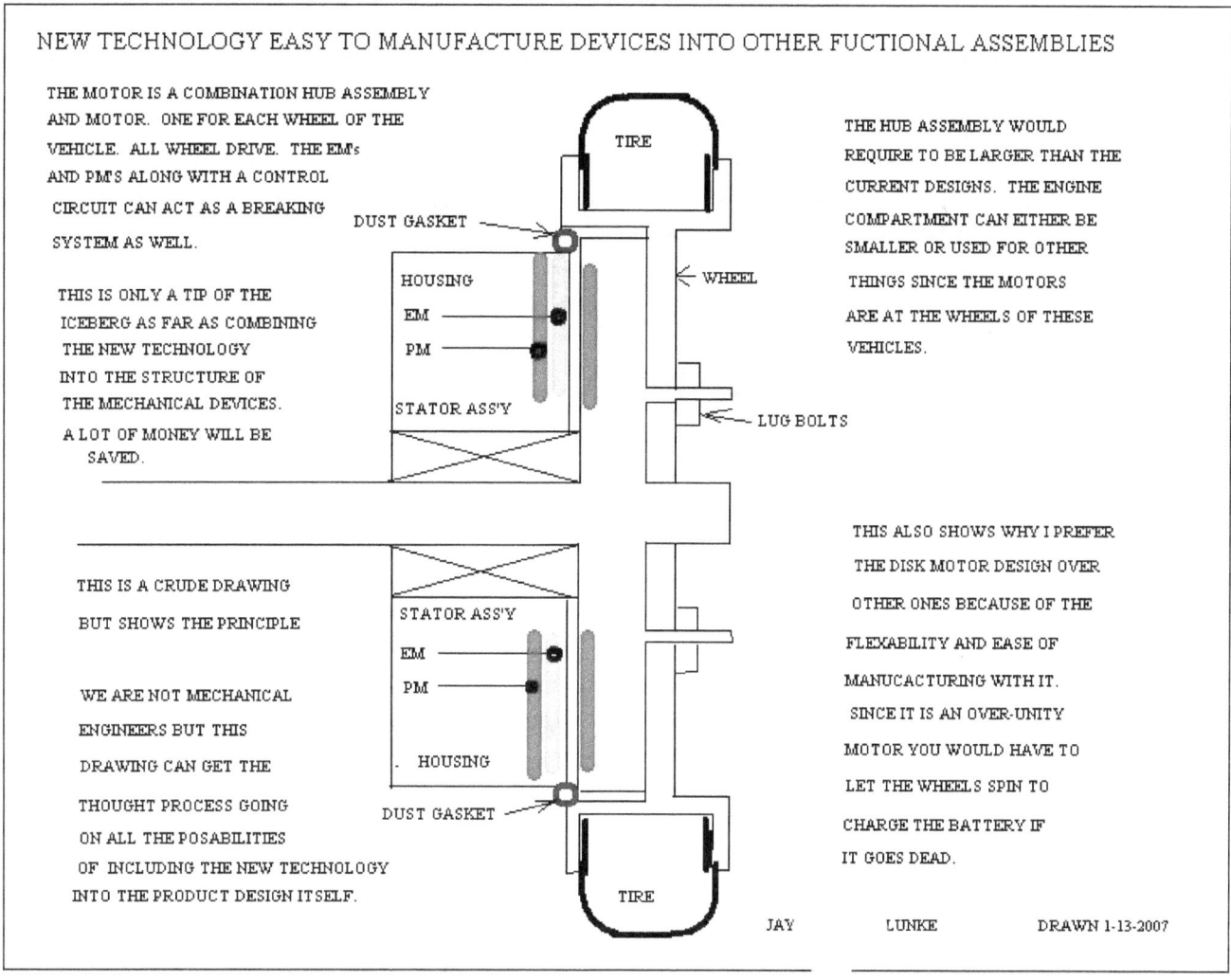

NEW TECHNOLOGY EASY TO MANUFACTURE DEVICES INTO OTHER FUCTIONAL ASSEMBLIES

THE MOTOR IS A COMBINATION HUB ASSEMBLY AND MOTOR. ONE FOR EACH WHEEL OF THE VEHICLE. ALL WHEEL DRIVE. THE EM's AND PM's ALONG WITH A CONTROL CIRCUIT CAN ACT AS A BREAKING SYSTEM AS WELL.

THIS IS ONLY A TIP OF THE ICEBERG AS FAR AS COMBINING THE NEW TECHNOLOGY INTO THE STRUCTURE OF THE MECHANICAL DEVICES. A LOT OF MONEY WILL BE SAVED.

THIS IS A CRUDE DRAWING BUT SHOWS THE PRINCIPLE

WE ARE NOT MECHANICAL ENGINEERS BUT THIS DRAWING CAN GET THE THOUGHT PROCESS GOING ON ALL THE POSABILITIES OF INCLUDING THE NEW TECHNOLOGY INTO THE PRODUCT DESIGN ITSELF.

DUST GASKET

HOUSING

EM

PM

STATOR ASS'Y

TIRE

WHEEL

LUG BOLTS

STATOR ASS'Y

EM

PM

HOUSING

DUST GASKET

TIRE

THE HUB ASSEMBLY WOULD REQUIRE TO BE LARGER THAN THE CURRENT DESIGNS. THE ENGINE COMPARTMENT CAN EITHER BE SMALLER OR USED FOR OTHER THINGS SINCE THE MOTORS ARE AT THE WHEELS OF THESE VEHICLES.

THIS ALSO SHOWS WHY I PREFER THE DISK MOTOR DESIGN OVER OTHER ONES BECAUSE OF THE FLEXABILITY AND EASE OF MANUCACTURING WITH IT. SINCE IT IS AN OVER-UNITY MOTOR YOU WOULD HAVE TO LET THE WHEELS SPIN TO CHARGE THE BATTERY IF IT GOES DEAD.

JAY   LUNKE   DRAWN 1-13-2007

I have many more drawings and sketches of different motor configurations and applications that I have not shown in this paper or the old paper because it would convolute my papers.

HOW DOES THE MOTOR
WORK AND NOT CONFLICT
WITH PHYSICS?

It starts at the atomic level of electron movement. Every atom has electron flow around the nucleus. No matter how many times the atom is moved throughout eternity, the electrons still move around the nucleus. Now show me a case in the normal usage of atoms where the electrons do not flow around the nucleus.

Now let's look at the permanent magnet. It takes the random flow that are in a group of atoms and lines them up so that the orbits are all moving in the same direction. This alignment creates a magnetic field where the flux flows out of one end of the magnet and comes back through the other end of the magnet. While the magnetic force(flux) is moving outside the magnet to go back to the other side of the magnet, it picks the easiest flow route which is not always the shortest route. It is the flux switch that determines the route that the flux will travel. The flux still has its physics of reactions in the route of the travel it is in. So, the timing and re-routing of the magnetic flux interactions with the other motor component is what generates the forces to keep the motor rotating. As long as the atoms in the permanent magnet are lined up with each other, then the magnetic power of the permanent magnet will continue to occur in the magnet. Physics has been used in the new technology not ignored. The big question is this. Is the electrical energy used for the switching going to be the same as that of a conventional electric motor? Can mechanical switching be demonstrated in one of these motor designs?

The electrical power control circuit could reduce the motors electrical power needs for this motor. Now read any literature of the operation and efficiency of tank circuits used in electronic circuits. By modifying this simple circuit to do two things will even more greatly improve the efficiencies of this motor. The fact that a lot of the electrical energy used to create a magnetic field can be re-captured and used again is not new. Work is being done in many places to study and develop this approach in motor assemblies. A lot of them try to capture the flux using other coils and then reusing the energy. In my approach with this new technology, the same coil that created the flux in the first place is used to capture this energy. After a magnetic field is created by the electro-magnet, like in all electro-magnets, when the power is turned off from them, a back EMF is generated. It just so happens that the physical position of my motor uses this for additional push on the rotor assembly. On most electro-magnets, the back EMF is waisted energy, never to be used again. By using the tank circuit, I can capture some of that energy back in my circuit while the back EMF is occurring. This is captured into the capacitor of the tank circuit. This captured energy is then used again in the electro-magnet is turned on. The electro-magnet is powered in the second segment when the electro-magnet has the greatest torque from the tank circuit. When an electro-magnet is turned off it discharges and produces a back EMF. So, I take advantage of it in my motor designs. The torque produced in the electro-magnets is smaller when the rotor is moving through the following segment of travel. I call the EMF a free push when I write about it.

The variable capacitor in the circuit helps to optimize the circuit while the motor undergoes varying load conditions. If you have read the efficiencies of the tank circuit, you know that this is the best approach for the operation of this new technology.

When you examine the losses of the conventional induction motor closely, you will not find that the magnetic torque is not a part of that equation. They have core losses, eddy current losses, wire winding resistance losses. The motor efficiencies are calculated by subtracting these losses // from the motors potential output. This would also mean that the mechanical power of a generator has similar losses in calculating the efficiencies of the generator. What this means is that an electric motor operating with all electro-magnets, using its torque output to operate a generator to create electricity could never reach 100%. This is because you could never eliminate all the losses in both the motor and the generator.

When using a power rotor, the new technology uses three magnetic torque powers and two electro-magnets' worth of power used in segment two. When using an efficiency rotor, the new technology uses three magnetic torque powers and one electro-magnet's worth of power in the functional stator magnet that is created in segment two. The electrical energy is less because of this. This motor operates more efficiently than conventional motors because the losses are less with the smaller duty cycle used in the motor's movement. The torque per watt in my motors are greater than the electrical energy used in conventional motors because I use a ratio of up to four to one, energy from the permanent magnets compared to the electrical energy from the electro-magnets.

WHY REDUCE THE ELECTRO-MAGNETS IN THIS NEW TECHNOLOGY?

It is because the permanent magnet is like the sun, wind or running water. The magnetic flux in a permanent magnet can be tapped into and used to create physical movement. The movement of electrons moving around atoms in your hand are what keeps your hand together. Being able to align this movement into a permanent magnet is a power pack to be tapped into. The flux switch is that tapping device to access this power into mechanical movement. By using the flux switch to control the interactions of the forces in the permanent magnet is what brings another option to the table like the usage of the sun, wind and water, to power our manmade devices. The efficiencies will be the highest of electric motors on the market so far.

NOW WHAT!

The next thing is to seek the best power circuit with low electrical losses to the system. I need to capture what power I can from the electro-magnets and then reuse it in the circuit again. I need to do this at the resonant point of the electromagnet to reduce the losses in the motor. I think that with the proper circuit, I could capture some of the electrical energy used in the electro-magnet to create the magnetic field back into the capacitor of the tank circuit. The amount of electrical energy captured and reused in the motor will increase the efficiency of the motor assembly.

I do not see anything in the physics for calculating electrical circuits and motor circuits that this new technology violates. The circuit efficiency will always calculate to less than 100%. But the efficiency calculations are not used in the torque output of the motor. The torque improvements in the motor because of the ratio of permanent magnets to electro-magnets and improvements of electrical energy due to the new power circuitry will produce an outstanding motor to reach the market for years to come.

FLUX SWITCH OPTIONS IN DETAIL:

Replacing the electro-magnet in the second segment of travel with either a revolving core or rotating permanent magnet or inserting a core or magnet in and out of the flux switching position: Since we can operate either a power rotor or an efficiency rotor with mechanical switching, I will concentrate the talk around using the efficiency rotor since when going to the extreme of mechanical switching, the goal would be for motor overall efficiency to be obtained.

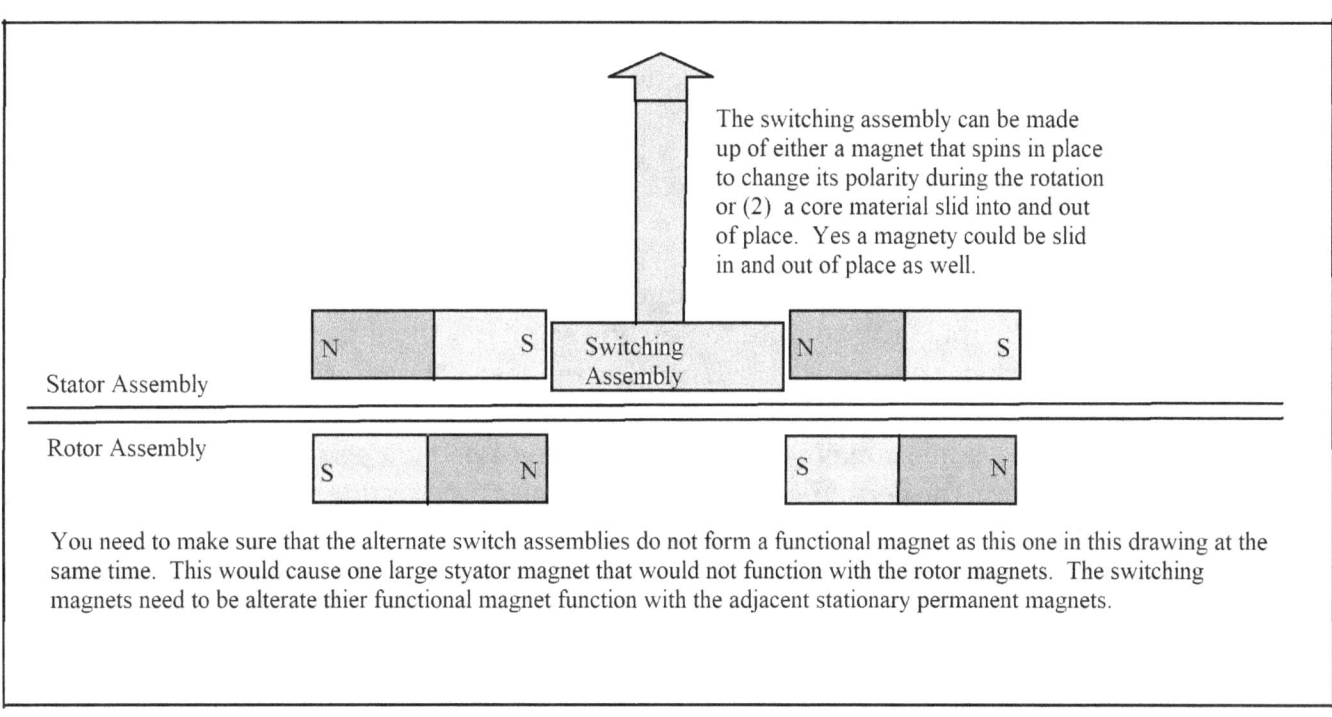

The switching assembly can be made up of either a magnet that spins in place to change its polarity during the rotation or (2) a core material slid into and out of place. Yes a magnety could be slid in and out of place as well.

Stator Assembly

N   S   Switching Assembly   N   S

Rotor Assembly

S   N   S   N

You need to make sure that the alternate switch assemblies do not form a functional magnet as this one in this drawing at the same time. This would cause one large styator magnet that would not function with the rotor magnets. The switching magnets need to be alterate thier functional magnet function with the adjacent stationary permanent magnets.

The segment one travel of the motor will have the stator permanent magnet attraction to the rotor permanent magnets as was the case with the P/P segment movement when an electro-magnet was in the motor with the power off. This will have one torque of power coming from the stator magnet and one from the rotor magnet.

The second segment of travel changes. In this configuration, either a core is brought into the mechanical circuit in order to create the ring magnet of the stator permanent magnets during this second segment of travel. Then the core is pulled out of the position of the stator magnets. This is done at the end of segment two's rotor travel. The mechanical movement is designed so that the core becomes active with the stator magnets at the position that the rotor enters the second segment of travel. The core is revolved out at the end of the second segment of rotor travel. The force of attraction from the permanent magnets to pull the core into alignment with the stator permanent magnets should be the same amount of force of repulsion needed to move the core out of range of the stator permanent magnets. So, over the operating range of the motor, these two forces should nullify each other out. The rotor should have small losses during the second segment of travel because it has little to no interaction with the stator permanent magnets during this time. The efficiency mode means that every other flux switch in the motor assembly is put into place creating functional stator magnets that will interact with the rotor magnets to create forward torque on the rotor assembly.

Another option for the power rotor is to have a permanent magnet added into the stator that is rotated so that when the rotor is in segment one, the magnet repels the magnetic flux of the stator magnets. When the rotor rotates into the second segment, the magnet is rotated to attract the flux in the stator magnets to generate the large ring magnet. The rotating permanent magnet would have the same 50% of attraction and 50% of repulsion of moving the core into and out of the range of the stator permanent magnets. The advantage of using a permanent magnet here is that it is a lot easier to rotate the permanent magnet in a circle than to slide the core into and out of position with the stator magnets. If the switching magnet has twice the flux as the stator magnet, then the additional flux would interact with the rotor in moving it through the second segment of travel. The down side is that the force needed to move the magnet would be close or the same as the forward torque on the rotor nullifying the added flux into the circuit.

The total motor will have a 50% duty cycle of torque from it with no outside force upon it. Do you know what this means? It means at least one more stator assembly should be added to create on the other side of the rotor to build a motor package. It would need to be installed out of physical phase with the first stator assembly in order to create torque through 360 degrees of the motor package's rotation. The current teachings of physics teach that it will take more energy to rotate the magnet into place or move the core in and out of place than the power torque produced in the first segment of motor movement. This is too bad because having no need for an electrical operating system would be enormous. I have the mechanical switching in place because I can think of some options where this type of motor configuration can be used.

Since more than two segments occur to complete 360 degrees of rotation, the simple main cam shaft approach will not be the optimal way for bringing the core in and out of position with the stator magnets. One design would be to bring the core in perpendicular into position between the stator permanent magnets. A large gear on the motor shaft could drive the smaller gears on the housing assembly that push the cores in and out of position with the stator permanent magnets.

Another option that improves the single stator motor assembly is to use the disk configuration for the motor assembly. The permanent magnets already go through the thickness of the rotor assembly for the power rotor. For the efficiency rotor, the magnets are more perpendicular to the stator magnets making the rotor wider to accommodate this difference. So, I can use both sides of that disk or drum to create the three-layered technology to work on them. There would be two stator assemblies, one on the left side of the rotor and one on the right side of the rotor assembly. This would bring the torque on the Power Rotor design back up to 360 degrees of the rotation. The rotor permanent magnet to stator permanent magnet action on the left side of the rotor occurs in the first segment. The rotor permanent magnet to stator permanent magnet action on the right side of the rotor occurs in the second segment. This whole process of two segment travel repeats itself again. The net result is 360 degrees of forward torque from the motor package.

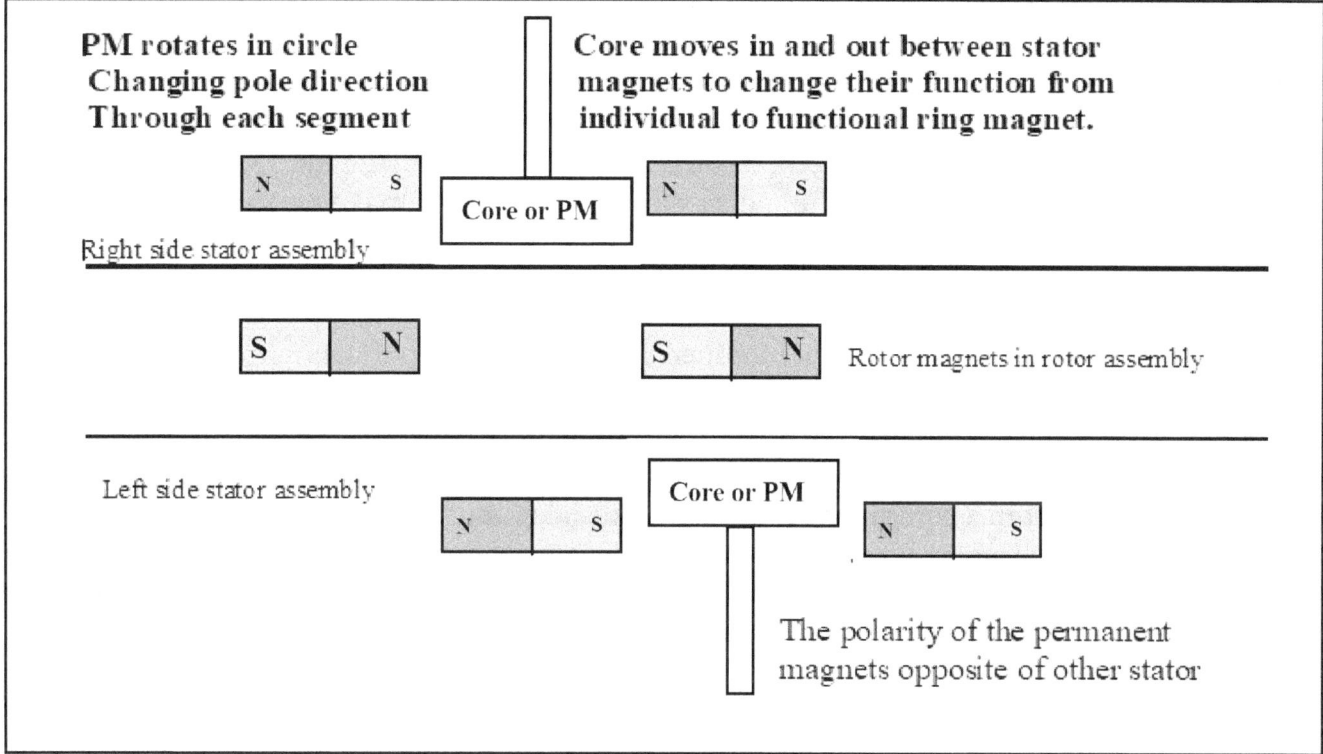

The Motor/Generator Combination:

The motor could have a generator built with part of the assembly on the rotor and part on the stator assembly. The rotor could be built wider like in a drum assembly. The reason for this is so that the generator permanent magnets should not interfere with the motor permanent magnets. To begin with I would build the permanent magnets of the generator at about the same distance from the motor shaft so they are moving at the same speed as each other. I would have the core material and permanent magnets configured in a closed flux loop with an opening between the North and South poles large enough for the stator electro-magnets to flow through. This system well defines the flux lines in the generator, making a more efficient transfer of flux lines into electrical energy in the electrical circuit. This energy is then fed to the motor electro-magnet to operate the motor assembly. Each coil in the generator feeds one coil in the motor assembly to provide the 25% duty cycle of the efficient rotor motor configuration. The ideal signal to drive the motor electro-magnet, would be a square wave the length of time for the rotor to travel through one segment of travel. Since I cannot achieve this with the generator I am building on my rotor and stator assembly, I will design in the following features. I will have a shorter coil in the generator along with a shorter permanent magnet assembly so that the pickup coil of the generator starts entering into a shorted permanent before the start of the segment length, and completely leaving the permanent magnets after the segment travel. I will start the segment travel when the pickup coil is about ½ ways into the permanent magnets and end the travel when the pickup coil is about ½ ways out of the permanent magnet. The reason I am doing this is because I want to make sure the rotor always has torque on it when the motor is running. If the power starts to early or ends too late, it does not create repulsion of the system, it at most reduces the torque of the motor assembly. In this way, the power in the electro-magnets will be about ½ of the full potential that it will generate when it is fully inside of the flux lines of the permanent magnet. By having the electro-magnet shorter in length than the permanent magnet, will cause a constant current fed to the electro-magnet during this time. By adding more layers of windings in the shorter pickup coil will make up the difference in the needed current for the motor's electro-magnets. I would actually want to use the same wire gage in the generator and motor electro-magnets. I would even design the pickup coils to be stronger than the motor electro-magnets to make up for the copper wire resistance and other losses in the circuit.

The pick-up coil will be made up in a modified of functional toroid configuration because the back EMF does not create a repulsion that reduces torque on the motors output. The core of the toroid could be built in a rectangular shape in order to meet the design criteria listed above. The permanent magnets that are on the rotor to produce the flux lines for the toroid coil to move through will be built from a combination of core material and permanent magnets into a square shape with an opening in one side for the toroid coils to move through. The reason for this is that I want to have well defined flux lines flowing for the toroid pickup coil to flow through so that the generator is as efficient as possible by reducing any means of reductions to the motor's performance.

The magical question here is will the electrical energy generated in the generator be enough to drive the motor circuit to be able to sustain its movement? The generator cannot generate enough electricity to drive the second stage by itself because of the losses in the electrical circuit. But remember that the first segment of travel did not require any energy outside of the internal magnetic forces inside of the permanent magnets to create the torque on the rotor assembly. So, the first stage would need to over-come the less productive generator to feed the motor electro-magnets in the motor assembly creating the torque on the motor assembly. Since the assembly of this circuit is easier than building a motor control and switching circuitry for this motor design, then I say why not build it into the proto-type motor to begin with. If it does not work, then I can use the pickup coils for signals to control the switching of the motor electro-magnets. If it does work, then I already have a demonstratable proto-type to show the world.

Now the motor with an on-board generator will not run without any electrical energy going through the electro-magnets. So, this design would need a push start. It also would need and way to control the speed of the motor. It would also need a way to stop the motor. The easiest way to do this is to place three or four-way switches between the pickup coils in the generator and the electro-magnets of the motor assembly. Also, in the switching would be a power supply control circuit. The switches would be switched to having the power supply circuit connected to the electro-magnets during the startup of the motor. After the motor got up to speed, then the switches would have the pickup coils switched to the motor to power the motor assembly. The switches could also be turned on and off individually in order to provide different steps of speed from the motor assembly. In order to completely stop the motor, you would simply open all of the switches.

The power control circuitry could generate power when breaking. This would increase the overall efficiency of the system. The applications would include car, trucks, motorcycles, bicycles, ext. Applications with a lot of stopping.

The thicker wire can be used between the generator and motor coils to reduce resistive losses in the system. The smaller the air gap in the motor, may improve the electro-magnet.

RESEARCH AND/OR DEVELOPMENT TO BE CARRIED OUT:

Plan A: **I may not get to build this prototype but maybe you can**

To Build a proto-type to prove the concept of the motor. I will take apart a bench grinder and gut it out. I would remove the very middle plate so that I can build the rotor drum assembly onto the shaft. The ends of the housing would be attached to aluminum plates larger than the discs and bolted together at the four corners to create stability. I would build two stator assemblies, one into each of the housing plates. I would have signal magnets on the rotor assembly mounted in the spacing between the rotor power magnets in order to signal the four segment power cycles that occur and repeat themselves. I could do this by having one signal magnet on the rotor and four pickup coils spaced out on the housing per every four segments of rotor travel. The motor will have two rotating outputs for evaluating the performance of the motor.

Plan B: **I may not get to build this prototype but maybe you can**

To build a proto-type of the motor that can easily be modified to different configurations. It would need to include the revolving core to test out the 100% mechanical motor configuration. This configuration should be tested first, because the others types may not be needed anymore. All the time I worked with other configurations would all but be gone, but they were learning experiences to bring me where I am today. I could then work on the application side of the technology.

Plan C; You could be the first to build this prototype

Build and testing of the disk motor assembly:

The motor assembly will be built up in module form in order to provide for assembly and reassembly into different design configurations for testing and evaluation purposes. I would build four disk assemblies. Two disk assemblies would be rotor assemblies. One would have permanent magnets installed every fourth segment on the disk assembly. The other rotor disk would have the permanent magnets installed every other segment location. Disk three will be the electro-magnet disk assembly. The electro-magnets will be placed at every other segment location on the disk. Each electro-magnet will have separate wiring so I can test different power schemes including 25% and 50% duty cycles for operating with the deferent rotor disk assemblies. The electro-magnet can be tested on either the rotor or the stator assembly. I will start with them installed on the stator assembly. Disk number four will be the stator disk assembly built up of permanent magnets placed in every other segment location. I call and often draw the stator assembly as a disk assembly when most of the time it is not shaped as a round disk. The reason I do this is because the magnets are laid out the same as the rotor disk assembly in order to interact with it. So, the rotor dictates the stator design in the area of the interaction of the two assemblies. The rest of the housing stator assembly has so many options in how to build it so I do not talk about it very much.

One should first test the performance of disk motor assembly without the stator permanent magnets and record the performance. This testing will simulate testing of a PM motor assembly that is efficient but will not be the best design for its efficiency ratings. One should then add the permanent disk assembly to the stator assembly to collect test data. This configuration will be testing the new technology I am promoting. The two sets of test data will be used to evaluate and compare the two designs. Basically, this would be comparing two-layer technology against three-layer technology. This testing will show how much the new technology adds to the performance in mechanical movement in mechanical devices.

Someone needs to optimize the Solenoid design. A need to design an electro-magnet that will operate with the most magnetic field strength when the power is turned on at the same time having the least effect on the permanent magnets when the power is turned off. There is a need to optimize the dimensions of the solenoid coil. These designs could be different for different design applications.

1.) Test out different wire gauges
2.) Test out different amounts of winding layers
3.) Test out taping the coil at different layers. It would be good to supply different voltages to them by connecting some in series and some in parallel
4.) Test the above tests using different length coils
5.) Test the above test using both different size cores and different core materials. One could build the coils around different size aluminum tubes so they could slide different core materials in them to accelerate assembly of different configurations to reduce the time to get the assemblies ready for the next test. This would save time in winding new coil assemblies around the cores each time one wanted to evaluate a new core material.

The distance between the stator assembly magnets and rotor magnets needs to be evaluated at the same time. This testing will be done along with the electro-magnet evaluation testing.

Evaluation of the optimal number of segments set in a disc should be evaluated.

Also, how many layers of PFPMMDs the motor should have in it.

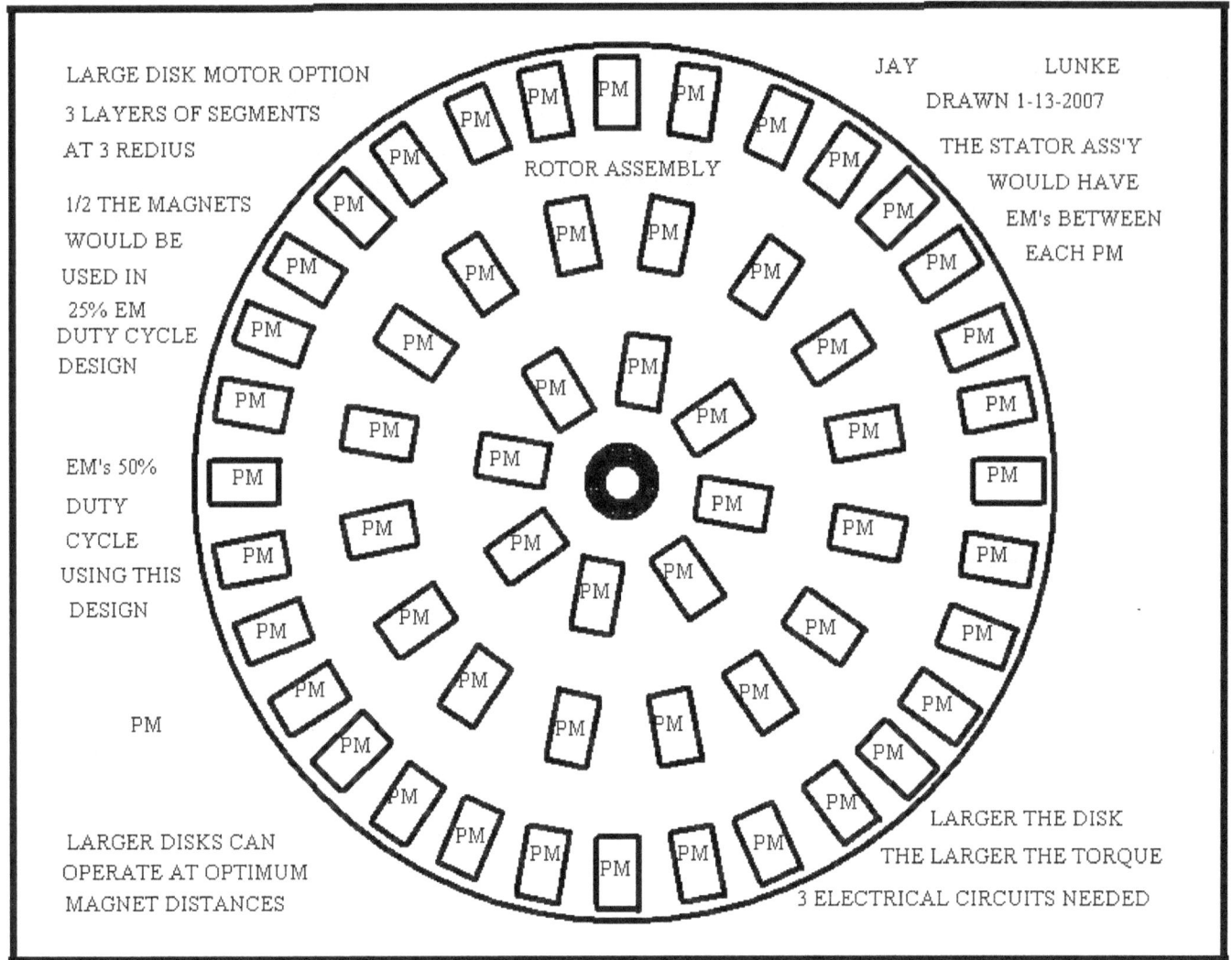

LARGE DISK MOTOR OPTION
3 LAYERS OF SEGMENTS
AT 3 REDIUS

1/2 THE MAGNETS
WOULD BE
USED IN
25% EM
DUTY CYCLE
DESIGN

EM's 50%
DUTY
CYCLE
USING THIS
DESIGN

PM

LARGER DISKS CAN
OPERATE AT OPTIMUM
MAGNET DISTANCES

ROTOR ASSEMBLY

JAY   LUNKE
DRAWN 1-13-2007

THE STATOR ASS'Y
WOULD HAVE
EM's BETWEEN
EACH PM

LARGER THE DISK
THE LARGER THE TORQUE
3 ELECTRICAL CIRCUITS NEEDED

The optimal mechanical assembly for performance objectives should be explored:

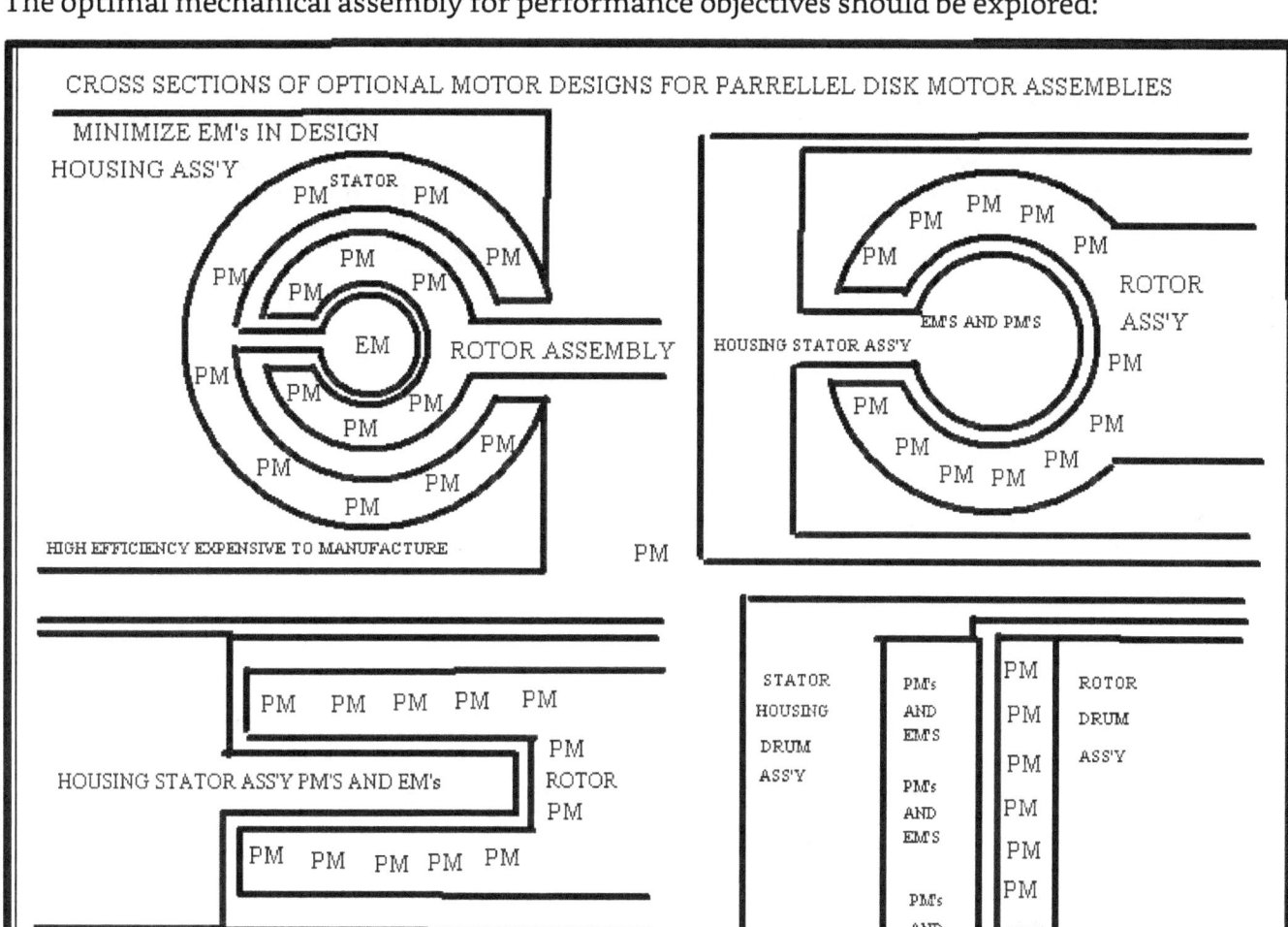

There are an unlimited design options that use the three-layered approach to electrical-mechanical movement. The purpose of this technology is to use more of the permanent magnets into the three-layer designs.

1. To take advantage of the permanent magnet to permanent magnet interaction with each other to create as much physical movement in the motor as possible.
2. To reduce the amount of the electro-magnets in creating the physical movement of the device.
3. To use the electro-magnets in their most efficient operation state by adding power/control circuitry like that of the resonant tank circuit.
4. To make it maintainable.
5. To make it manufacturable.
6. To incorporate them into the application hardware when possible.
7. To refine the components through optimization research methods.

How is the three-layer technology different from current technologies?

The conventional electro-mechanical electric motor operates functionally as a two-layer system. The movement in the device is made up of the magnet in the stator interacting with the magnet in the rotor to create the movement between them. Either the rotor magnet or the stator magnet is an electro-magnet. In order to operate the motor, the power to the electro-magnet is usually on all the time in order to keep torque on the motor at all times. Usually, if there is no power, there is no movement of the motor in any of its 360 degrees of rotation. The conventional motor operates with either permanent magnets or electro-magnets in the stator assembly.

Pulse motors also operate in a functional two-layer technology. Either the electro-magnet in the rotor is energized to interact with the magnet in the stator or the electro-magnet in the stator is energized to interact with the magnet in the rotor.

I have not done an extensive search of patented electric motors yet because I did not want the ideas for my technology to stem from another designer's work.

Background of the invention or discovery, testing, observation, theorizing, etc.
I started working with magnetic movement starting in 1969 by putting two sets of permanent magnets in a track assembly together with a permanent magnet in the middle to represent the electro-magnets on a device.

I picked up the work on a new electric motor called the flow through electric motor in 1982 for one year. These motor designs are high power, high torque motors using a mass ratio of about 10 to 1 of permanent magnets to electro-magnets. These motors require switched polarity electrical energy all the time on the electro-magnets. I designed several optional electrical driver circuitries including a circuit that would operate the motor in a tank circuit with the coils in the motor assembly as part of the tank circuit which would allow the motor to operate at its resonant point at different motor speeds and loads by adjusting the capacitor value in the tank circuit on the fly as the motor is running. This circuit with a few modifications could be used to operate the new motor assembly as well.

I built a very crude proto-type motor to prove the motors functionality. It was so crude, that I destroyed it, I did not have the funds to build a full proto-type for outside Lab testing so I stopped working on my motors until January of 2006. This new motor was more of a long shot for success over other motors that were being built with high efficiency, so I did not feel that I could compete with them with as low of funds that I had at the time. Some of the motors claimed over-unity gain. How can I compete with that?

I have since thought that my current designs have more of a chance at over-unity than theirs do when I use a combination of the power circuit with the new motor's designs.

But for years now I have not seen a big market change for those motor designs, maybe there is a problem with those designs. So, I want to see how efficient my new technology designs can be and if they can make a dent in the current market place.

It was in January of 2006 that I built a small test lab for the purpose of experimenting with different design concepts I have had. It was on this test station I built a sub-assembly with one permanent magnet on the rotor assembly and two permanent magnets on the stator assembly with one electro-magnet between them. I used a few different electro-magnet designs. I discovered that the assembly would function with as little as 100 milliamps. The sub-assembly did prove out the two-segment mechanical device movement producing positive torque on the rotor assembly in all positions of the motors travel. Part 1(P/P) requiring no external power because it is a permanent-to-permanent magnet reaction while the second movement is an efficient P.M / electro-magnet reaction. If I can get the same torque from Part 2(F/P) as the part 1(P/P) segment movement with an efficiency of even 50 percent in part 2(F/P) then the total efficiency of the motor assembly would be 100 percent. Now there could be less than a 50% efficiency in the part 2(F/P) segment movement due to overcoming the configuration of the permanent magnets interacting with the electro-magnet in the placement of the magnets in the assembly. Testing in a lab is the only way to answer these questions. I did not have the funds to purchase all the equipment to build a lab or to build an adequate prototype and pay to have a lab perform the testing on it.

I have been salvaging surplus and scrapped motors for parts for the test station and the prototype devices I have been working on.

Now with the work I have been doing, I do believe that over-unit being a bad term for this technology, is possible to achieve.

I had planned to make another simple proto-type to have complete rotation of the disk assemblies fully populated with components so that I can make a better assessment of the technology. I was building it out of wood and the wood warped and the prototype destroyed.

MECHANICAL MAKE UP OF PROTO-TYPE MOTOR ONE

Purpose: To create a flexible design that in itself holds many aspects of the future of electro-mechanical motors. In doing so, the design needs to consider the following features.

1.) The motor needs to address its own constraints as much as possible.

   A.) The large magnetic fields from the permanent magnets need to be protected. The permanent magnets need to be about 2 inches away from materials that magnetic fields are attracted to. This includes component mounting hardware and housing materials.

   B.) The Permanent magnets can be damaged in temperatures over 175 degrees. Proper heat sinking and air flow through the motor need to be incorporated in the design.

   C.) We do not want cross-talk occurring in the wiring. Each electro-magnet and its Toroid pick-up coils will be wired to their own separate connector with the shortest wire routing possible without compromising the servicing of the motor assembly.

   D.) FCC requirements may need to be addressed on the motor. The Toroid coils will greatly reduce signal transmitted noise, but depending on the RPM of the motor, the electro-magnets will likely create radiated signals that need to be addressed. Aluminum housings would add to the protection of these signals.

   E.) Higher rotor speeds will create large kinetic energy will mean that the rotor components will need extra protection. Having cutouts in the disc panels in the motor will provide a lot of that protection.

2.) I need to optimize interfacing with existing mechanical systems. Since I want to reduce the overall weight of my motors to begin with, I will remove the weight in the center of the motor by having the motor's rotor be built into a drum with the interior hollow. Both ends of the motor will accept an adaptor. There can be many adaptors with different shafts and other hardware on them so that the new motors can easily replace the old motors in their applications. The hollowed center goes through the whole motor. This will provide other options with the correct adaptors attached to the motor like pumping liquids through them.

3.) Having the permanent magnets in the stator able to be adjusted for the distance they are from the rotor magnets would be a nice to have option for the first go around of the motor.

4.) Having a little adjustability in the Toroid pickup coils would also be nice to have features in the motor.

5.) Having the connector type that brings the leads out of the motor at right angles would help cable control for the motor electrical control unit when it is built. I want to work with the proto-type when completing the details of the motor control unit.

6.) I want the design to be able to connect in additional motor modules into it as application needs require to do so. With the same attachment plates on each end of the motor, will allow another adaptor to connect another motor to the assembly. I need to make sure the drum is strong enough to add a couple more motors to it. Having adaptor mounting plates on each end provides several other applications to the motor.

7.) I found out through additional testing that the series parallel three-layer design will work better in this motor design when using the efficiency rotor configuration with it. It does not work in all attraction mode like the parallel configuration, but the repulsion between the stator and rotor permanent magnets does create forward torque on the motor assembly. Careful wording to include these optional configurations is important for the understanding of the motor designs for a wide range of readers reading them.

8.) This motor will have 24 segments on each side of the rotor. It will have two stator assemblies. The rotor will be a rotating drum with two sets of magnets in the rotor, one at each end of the drum. The center section of the drum will be the custom-made horseshoe style of permanent magnets in order to produce a constant flux field for the Toroid pickup coils to travel through in order either drive or signal the drive current in the electro-magnets. The coils and electro-magnets will be designed to be operated in a 24-volt system. The cores can be built out of transformer type of laminated materials. The distance between the rotor magnets and the pick up coild should not be more than 2 inches. There should be a minimum of about 1 inch between the two channel magnet assemblies used for the toroid magnets.

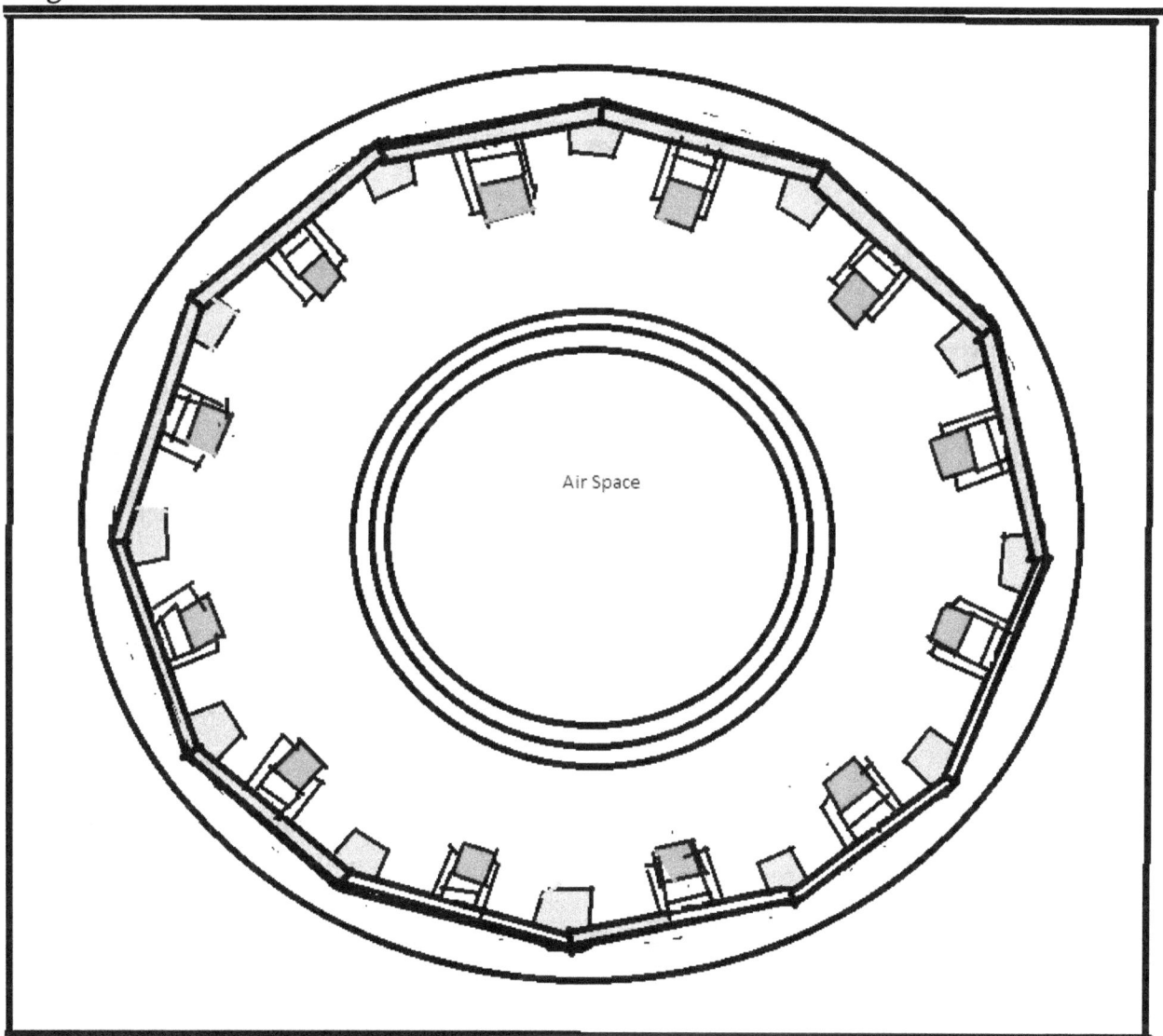

Air Space

Cross-section of one of the two Toroid pick up coils. The length of the coil is one half the length of the stator permanent magnet. The length of the flux channel that the coil moves through is also one half the length of the stator permanent magnet. When the coil first touches to flux channel, the rotor will have travels 3/4 of first segment travel. When the rotor starts into the second segment of travel, the coil will be 1/2 way into the flux field. When the coil is totally within the flux field, the rotor will have traveled to the center point of the second segment of travel.

If the stator magnets are 2 inches long, then the coild will be 1 inch long and the flux field will be 1 inch long. The permanent magnets can be built with the 1/2 inch cubs spred out in length and width to cover coil from top to bottom as it travels through the flux channel.

As you can see the sketches are not to spec. I do not show the connectors in this sketch either. The normal Toroid is normally round in shape, Thansformers are rectangular in shape with laminated cores.

The coils will be operating in a 24 volt system. The coil wire diameter is changed with voltage to give the same wattage output of the coil. I want the coils to be designed for a 30lb holding force with a constant 24 volts applied to them. This would also be true for the electro-magnet coils. This may not be achievable when the air core is used in them. If this is the case, then use the same wire gauge and wire length that was used to build the toroid coils.

Now if there are parts that are close to the same criteria on the market that can be found, then use them instead of custom parts. With a new motor technology, that my be hard to find.

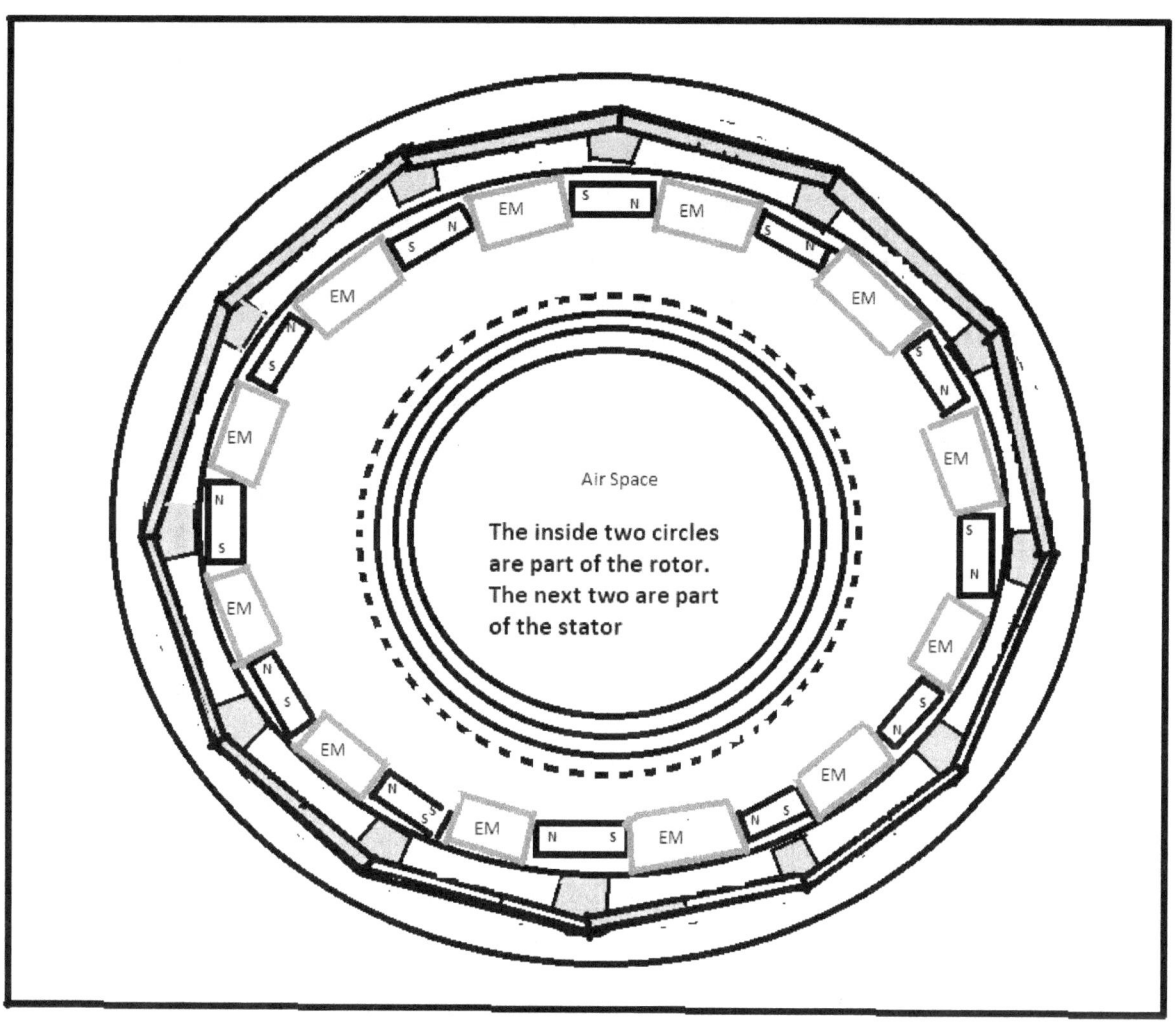

Air Space

The inside two circles are part of the rotor. The next two are part of the stator

This is only a sketch of the cross section of the stator assembly and housing to show the arrangement of the electro-magnets and the permanent magnets in the motor. I do not have measurements because there are different options that can be used for them. The segment width changes as you move out from the circumference. The magnet size will determine the size of the motor to some extent. There are several sizes to choose from. The most important thing about the magnets is the way the factory magnitised them. The placement of the poles of the permanent magnet are critical in all motor designs.

For this motor, 2 inch length permanent magnets is the best size, but 1.5" is about as short as I would want to design them. Being able to put spacers under the permanent magnets could be a design consideration. The permanent magnets in the rotor will be in every fourth segment with the magnets at a 90 degree angle to the stator permanent magnets. The rotor permanent magnets should be flush with the inside of the rotor so that they are as close to the rotor magnets as possible. No closer that the rotor to the electro-magnet s. The rotor magnets need to be at a minimum of 2 inches long.

I am not showing the connectors in this drawing. There will be 12 on each side of the motor

The following motor is similar but has only four pickup coils in the rotor assembly to signal the switching of power for the four segments of rotor travel. An external power circuit is needed to condition those signals and then drive the electro-magnets.

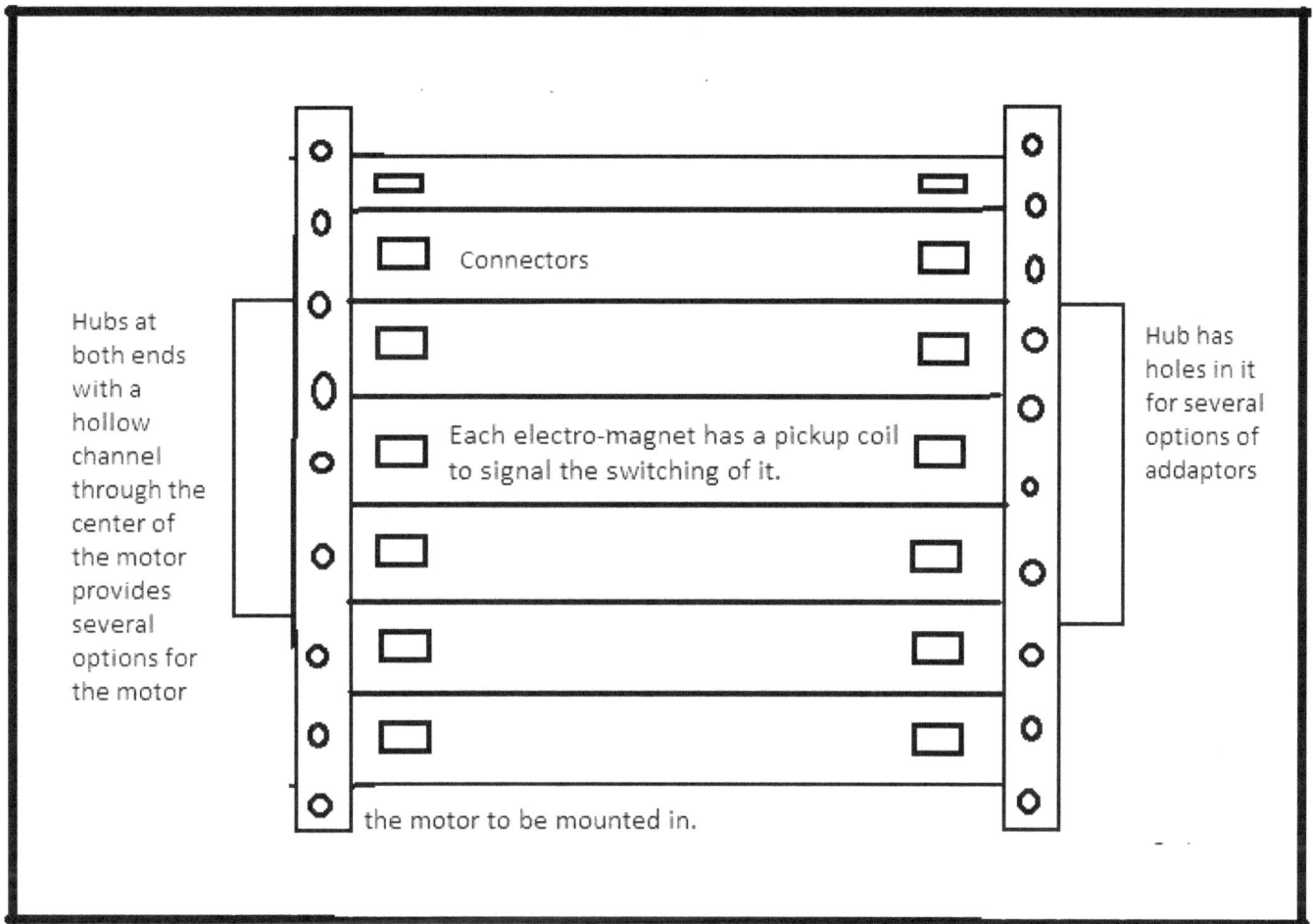

Hubs at both ends with a hollow channel through the center of the motor provides several options for the motor

Connectors

Each electro-magnet has a pickup coil to signal the switching of it.

Hub has holes in it for several options of addaptors

the motor to be mounted in.

Motor Assembly does not show air cooling vents or holes in this drawing.

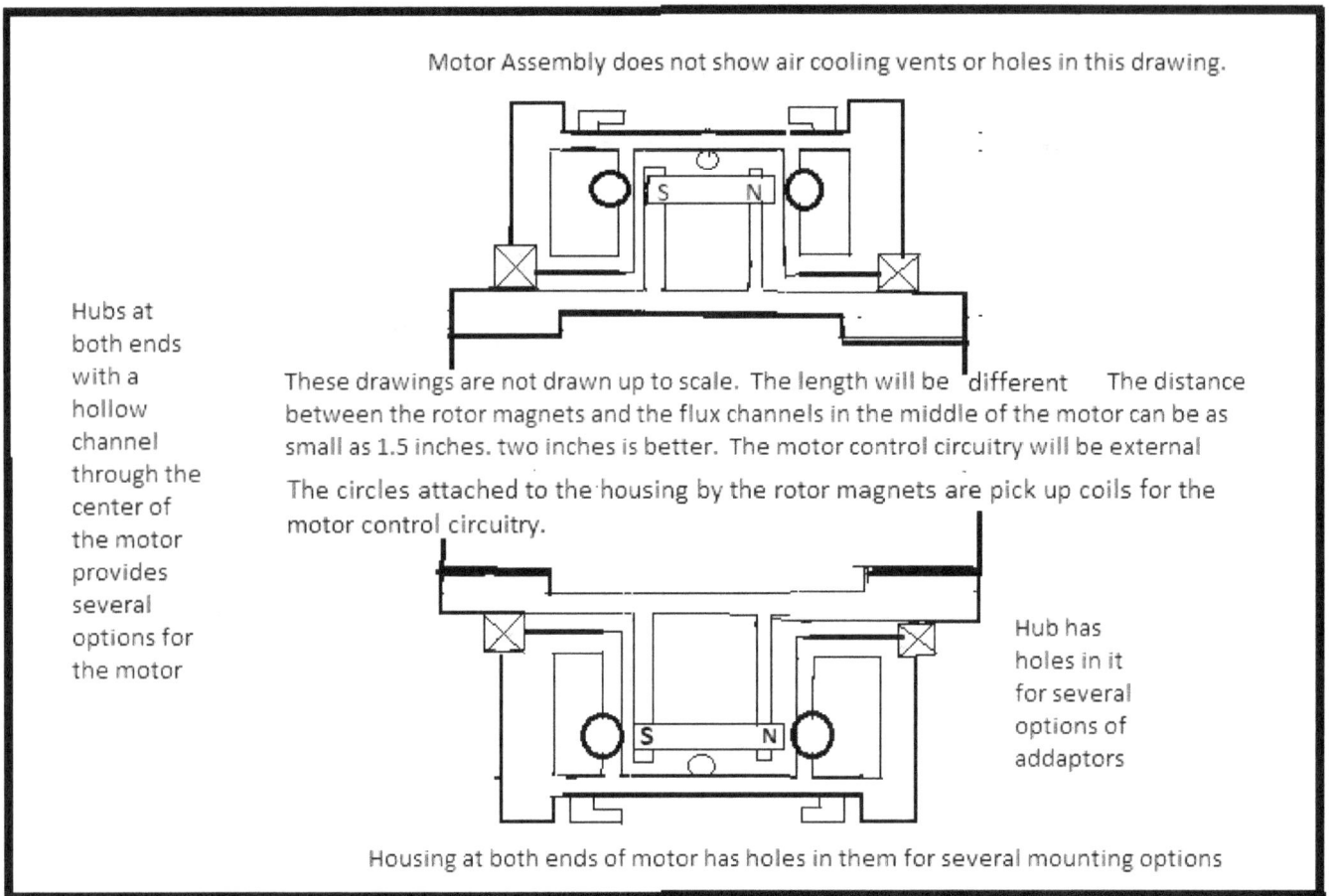

Hubs at both ends with a hollow channel through the center of the motor provides several options for the motor

These drawings are not drawn up to scale. The length will be different   The distance between the rotor magnets and the flux channels in the middle of the motor can be as small as 1.5 inches. two inches is better. The motor control circuitry will be external

The circles attached to the housing by the rotor magnets are pick up coils for the motor control circuitry.

Hub has holes in it for several options of addaptors

Housing at both ends of motor has holes in them for several mounting options

MULTI-PURPOSE ROTOR ASSEMBLY

There are times when you want to have the efficiency from a motor and there are times when you want the power from it. Both the power rotor and the efficiency rotor share the same stator design. The rotor could be designed so that the rotor could be converted from one style of rotor to the other on the fly by moving every other rotor magnet out of the path of the engagement area with the stator. This would work like a transmission on a car, only it occurs inside of the motor. The electrical circuit can easily be converted to operate both motor types. This would allow better performance in motor vehicles of both power and efficiency.

Another modification that should be made to motor vehicles is to be able to switch off the motor from the vehicle drive train to the generator. All three could be on the same shaft but it is the engagement and dis-engagement onto that shaft that needs to occur. After this happens, then the motor can be placed into the efficiency mode to run the generator, since the efficiency mode along with the Tank Circuit that reuses some of the energy operating the motor will likely be in a mode where the efficient generator could recharge the batteries and power the motor at the same time. There are a series of toroid generators that could be used for this purpose. Since you need a lot of power to move a vehicle, a full-time efficiency mode motor would not be able to produce the horse power needed to have the need performance for it. In the power mode, the batteries will be needed to run the motor. When you park the vehicle, then the efficiency mode will recharge the batteries in the vehicle without being plugged into an external power source. This will save the owner of the vehicle a lot of money over the life of the vehicle.

Of course, I do not have a working vehicle yet, but most inventions come to one's mind first. By freely giving these designs to the world, will allow many more people to improve upon what I have given to achieve my same goals of reducing the worlds usage of fossil fuels.

OLD WORK FOR MOST PEOPLE IS NEW WORK, SO HERE IT IS

FLOW THROUGH ELECTRO-MAGNETIC MOTOR: THEORY OF OPERATION:

It is a magnetic field flowing through another magnetic field. In order to do this, one of the two magnetic fields must be an electro-magnetic field. The magnet field that flows through the other magnetic field is the secondary field. The magnetic field that the secondary through is the primary field. As the secondary field approaches, moves through, and descends from the primary field, the electro-magnet current changes direction in order to change the polarity of the electro-magnet. Either or both, the primary and secondary fields can be built up with electro-magnets. In most of the motor configurations the secondary field will be the only one built up with electro-magnets.

In most applications, it is necessary to physically tap on to the secondary magnetic field to convert this transfer of energy to useable work. In order to tap this energy in a flow through motor, it is necessary to have a physical opening in the primary field's assembly. In most of the motor configurations the primary field is built up with permanent magnets. When tapping the secondary field, the most efficient movement is for the secondary field to move through the primary field in a straight line. For many applications this is not practical. For these applications the most common direction is in an arc. The smaller the curve of the arc, the more efficient and powerful the motor is. In order to expand the motors capabilities, both the primary and secondary fields can be built up with a group of magnets.

The most common route for the group of magnets traveling in the secondary field, is to travel in a series of arcs to add up to 360 degrees. And then start over the same path again. The electro-magnets can be mounted on a disc, drum, arm, assembly, housing or other mounting assembly. The most common mounting method is the disc assembly. The disc. Assembly requires the smallest opening in the primary field assembly to achieve the best performance characteristics.

Mechanical Characteristics:

Flow Through Motor Make Up

Type	Primary	Secondary
1.	Electro-magnet	Permanent Magnet
2.	Electro-magnet	Electro-magnet
3.	Permanent Magnet	Electro-magnet

Type 3 is the most commonly used.

The disc, electro-magnets, brush assemblies or contact strip assemblies, and other parts used in place of the armature is called the electro-magnetic disc assembly.

The distance that one electro-magnet travels to go through one permanent magnet is called one torque cycle. The number of torque cycles that a flow through electro-magnetic motor goes through for a 360-degree motor cycle is equal to the number of permanent magnets in the primary field of the type 3 motor. The permanent magnets used in a primary of a type 3 motor that function with an electro-magnet disc assemble, is called a permanent magnet ring. The cut-out area through the center of the permanent magnet ring is called the permanent magnet channel. See figure 1000 for details.

Unless otherwise told, I will be discussing the type number 3 motor style, operating in a circle.

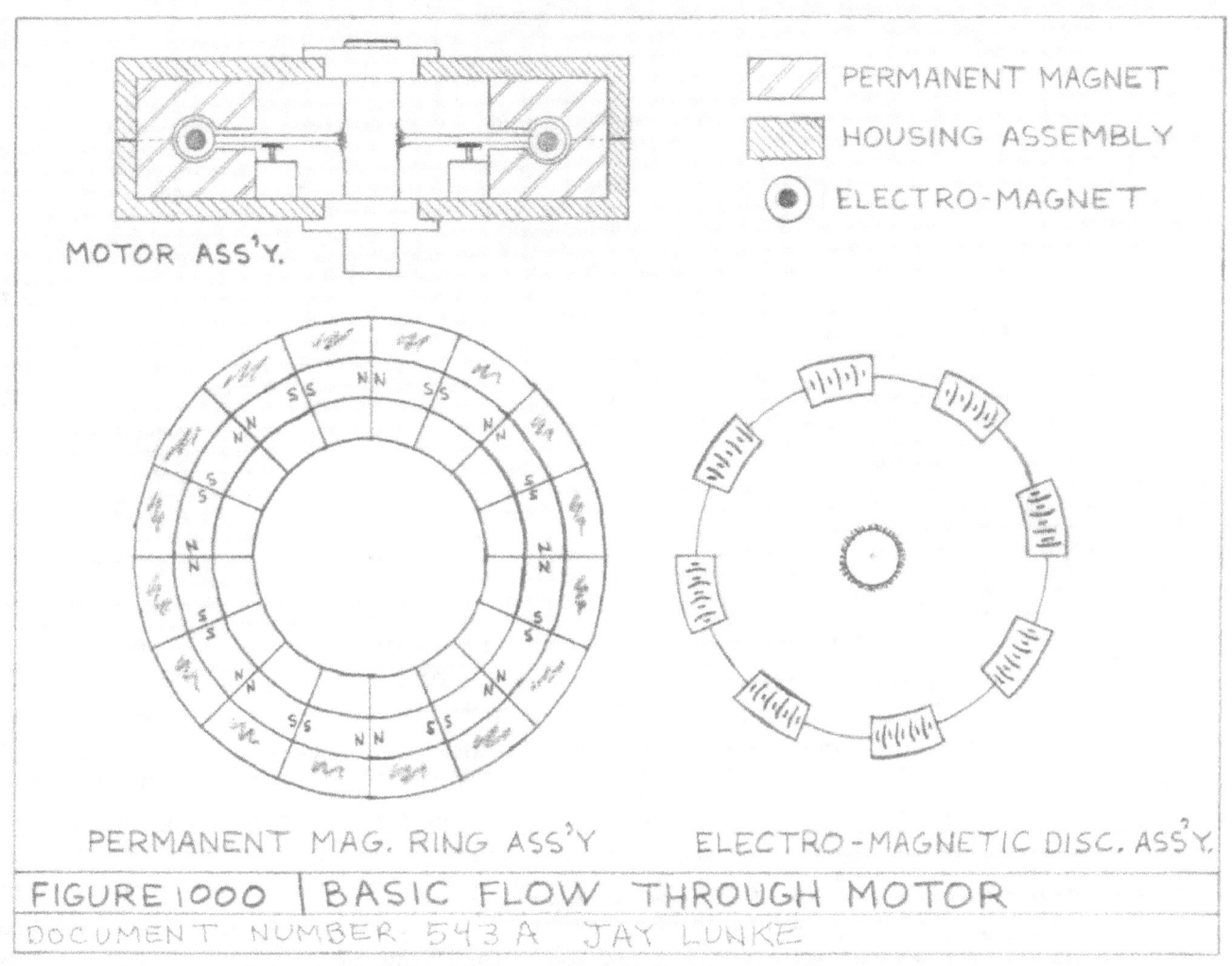

Comparisons to basic armature motors

(1.) The amount of area that the secondary field comes in direct or close contact to the primary field through a torque cycle is a lot greater in the flow through motor than the armature motor. As a result, the flow through motor can be built with less mass, or weight, to achieve the same power output. The armature motor has a lot of other assembly parts it comes in close contact with to promote losses. It also lowers the overall efficiency ratings of the motor. Most of the efficiency losses in the flow through motor will be in the electro-magnets themselves.

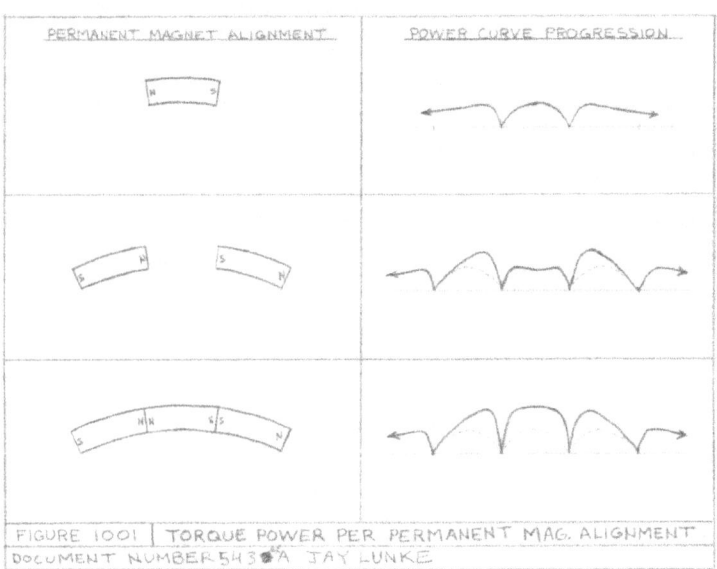

NOT TO SCALE:

(2.)   See figure 1001, this drawing shows the advantages of directly tying the magnets of the primary field together. The magnets must be tied together in such a way that each magnet in the magnetic ring repels the one next to it. This is done by installing like poles next to each other. The basic armature motor has two magnets and makes two torque cycles for a 360-degree rotation. One pole of each of the two magnets interact with the armature in both torque cycles. As seen in figure 1001, the characteristics of the power curve changes with the number and arrangement of the primary magnets. The biggest reason for this is because the electro-magnets are interacting with four magnetic poles instead of two. This means the flow through motor is capable of more power. In theory the power curve will not only be larger, but it will provide an almost square looking type of curve in comparison to the armature motor. This power curve not only indicates improvements for the power per pound ratio, but also increased motor efficiency. As shown in figure 1002, the advantages add together to make the flow through motor a very desirable motor design. The flow through motor has a much higher range of efficiency in comparison to the armature motor.

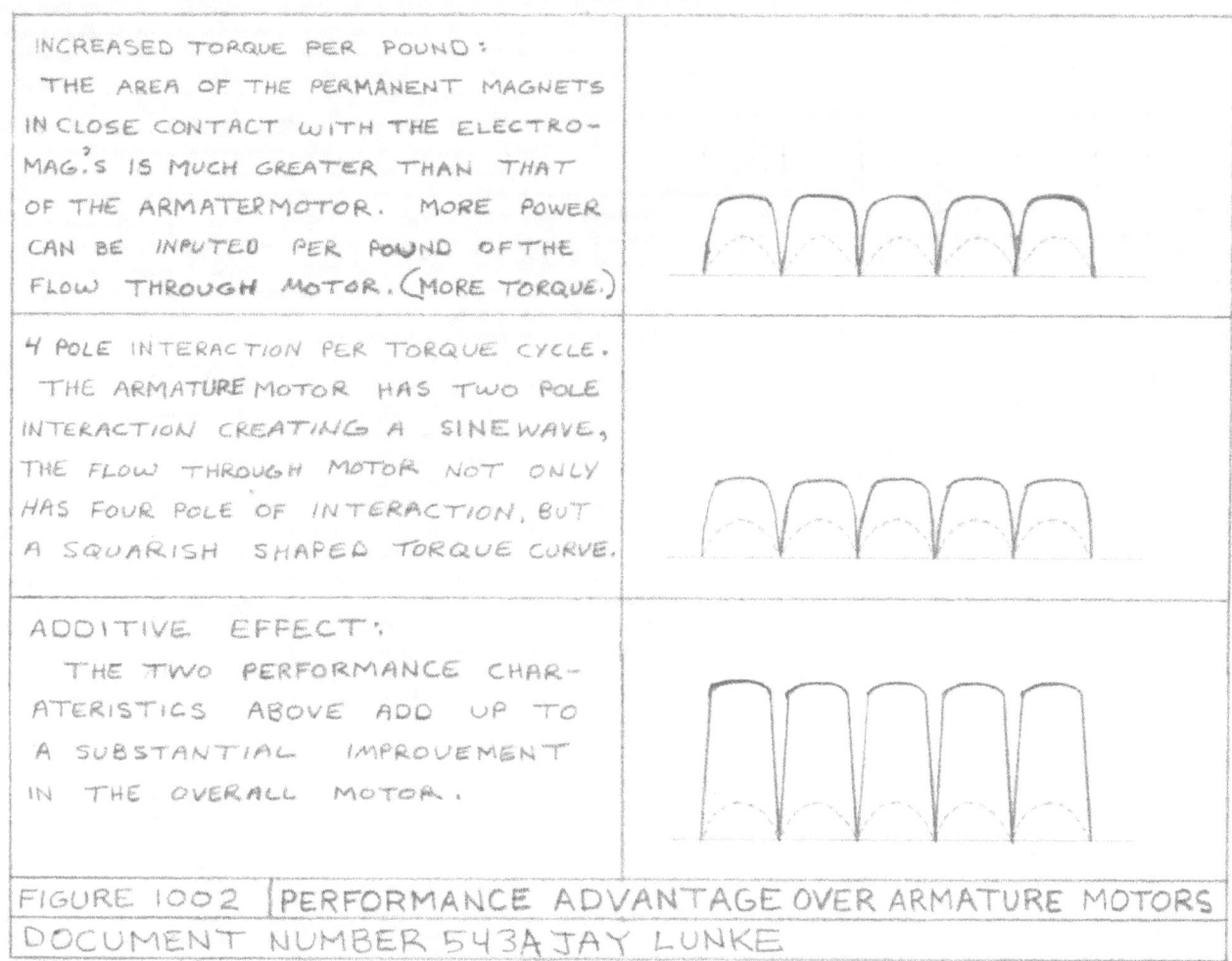

INCREASED TORQUE PER POUND:
THE AREA OF THE PERMANENT MAGNETS IN CLOSE CONTACT WITH THE ELECTRO-MAG.'S IS MUCH GREATER THAN THAT OF THE ARMATER MOTOR. MORE POWER CAN BE INPUTED PER POUND OF THE FLOW THROUGH MOTOR. (MORE TORQUE.)

4 POLE INTERACTION PER TORQUE CYCLE. THE ARMATURE MOTOR HAS TWO POLE INTERACTION CREATING A SINE WAVE, THE FLOW THROUGH MOTOR NOT ONLY HAS FOUR POLE OF INTERACTION, BUT A SQUARISH SHAPED TORQUE CURVE.

ADDITIVE EFFECT:
THE TWO PERFORMANCE CHAR-ATERISTICS ABOVE ADD UP TO A SUBSTANTIAL IMPROVEMENT IN THE OVERALL MOTOR.

FIGURE 1002 | PERFORMANCE ADVANTAGE OVER ARMATURE MOTORS
DOCUMENT NUMBER 543A JAY LUNKE

NOT TO SCALE:

(3.)   The diameter of the armature is limited in size, in a large part due to the armature's operational function. Since the electro-magnets do not travel through the center point of the motor, the shape and size of the electro-magnetic disc assembly is very variable in design shape. In fact, it is much easier to build and work with electro-magnet disc assemblies.

The larger the diameter of the disc is, the greater the torque of the motor becomes. The torque of the motor is not only dependent on the size of the diameter of the disc, but also on the number of permanent magnets in the permanent magnet ring. Figure 1003 shows both a torque diameter ratio chart.

TORQUE DIAMETER RATIO CHART : REFERENCE 10" DIA. SAME POWER INPUT

DIAMETER OF DISC.	TORQUE RATIO
10	1.0
20	2.0
30	3.0
40	4.0
50	5.0
60	6.0
70	7.0
80	8.0
90	9.0
100	10.0

THE FOLLOWING ADD UP FOR IMPROVED MOTOR PERFORMANCE:
1.) INCREASED TORQUE CYCLES.
2.) INCREASED DIAMETER.
3.) INCREASED PWR./LB.: PHYSICAL ARRANGEMENT.
4.) 4 POLE INTERACTION EACH TORQUE CYCLE.
5.) POWER CIRCUITS EFFICIENCIES.

TAILORED TORQUE CHART

# of Mag's in Magnetic Ring	Degrees Per Torque Cycle	Multiplier factor of torque advantage
4	88.95	2
6	60	3
8	44.875	4
10	36	5
12	30.	6
14	25.714	7
16	22.438	8
18	20.	9
20	18.	10
22	16.364	11
24	15.	12
26	13.846	13
28	12.857	14
30	12.	15
32	11.219	16

FIGURE 1003	PERFORMANCE CHARTS

DOCUMENT NUMBER 543A JAY LUNKE

The power to drive the motor increases linearly with the number of magnets:

(4.)   The flow through electric motors is very versatile for both its physical design and applications.

Multiple Magnetic Flow Through Motors:

Permanent Magnet Ring;

The permanent magnet ring does not have to be filled with permanent magnets. The permanent magnet ring can have as few as one magnet or have up to several permanent magnets for each permanent magnet. Most permanent magnet rings are completely full with permanent magnets. This style provides the power per size. Another option is to have the permanent magnet ring 50 percent full of permanent magnets. The length of the space between permanent magnets would be the same as the length of the permanent magnet. The advantages of this type of permanent magnet ring are as follows:

1. The permanent magnets are not in the physical position or polarity to be deteriorated over time from the strain that a full ring puts on a permanent magnet. The permanent magnets have their magnetic fields in the same direction in the permanent magnet channel.

2. The power circuitry required to operate the motor can be reduced in size.

The disadvantage of this type of permanent magnet ring is that the power output is less for the physical size and weight of the motor assembly.

Electro-magnetic Disc Assembly;

Like the permanent magnet ring, the electro-magnetic disc assembly can be built with as few as one electro-magnet to a disc having several electro-magnets, but the efficiency of the motor would be low because you would have losses due to the electro-magnets fighting each other. The best set up is for the electro-magnets to fill 50 percent of the disc that flows through the permanent magnet channel. The space between each electro-magnetic Disc Assembly Through a Permanent Magnet Ring Assemble:

When the physical position of the electro-magnet is lined up with the permanent magnet, they are at a null point. The motor is very inefficient around this point. It is at this point where the switching of power to the electro-magnets occurs. This causes the polarity of the electro-magnet to change direction. There is one null point per permanent magnet and the electro-magnetic field will change its polarity at each one of them. Except for the null point, the electro-magnet will always be leaving one magnet at the same time as it is entering another one. The nice part about this is that the electro-magnetic field that repels the electro-magnet from the permanent magnet it is leaving is the same force that attracts the permanent magnet it is moving into. This means that both the permanent magnets provide positive torque in moving the electro-magnetic disc assembly. After the electro-magnet reaches the next null point, the electro-magnet's polarity is changed. The electro-magnet has now moved to a position where the polarity of the permanent magnets has changed. Here again the interaction between the electro-magnet and both permanents provide positive torque. The whole cycle starts all over again.

The closer the electro-magnets can travel to the permanent magnets, the larger the power potential will be.

Many types of power circuits are available to operate and control the motors. These circuits are available to operate and control the motors. These circuits will be discussed later.

Here is a list of some of the advantages of the flow through motor over the basic armature motor;

1. Larger diameters provide larger torque outputs.
2. There is more torque potential and it is proportional to the number of permanent magnets.
3. With a full permanent magnet ring, you will get more torque output per magnet.
4. The motor lends itself to many design arrangements and applications.
5. The motor operates more efficiently due to the shape of the torque power curve.
6. The motor can produce a lot more power per pound.
7. The motor has a large range of options as to the kind of power circuits that will operate the motor.
8. Motor has lower heat generation.
9. The motor can be designed to operate at very high speeds.
10. The motor can be switched to operate as a more efficient generator.
11. It is easier to provide the option of using super conductors in the motor circuit.

12. The motor is very safe for both man and the environment.
13. It is a very low maintenance motor.
14. The motor has a good longevity. Except for the strain on the permanent magnets.

Permanent Magnets;

Figure 1004 shows some of the possible permanent magnet configurations that can be used to build the permanent magnet ring assemblies.

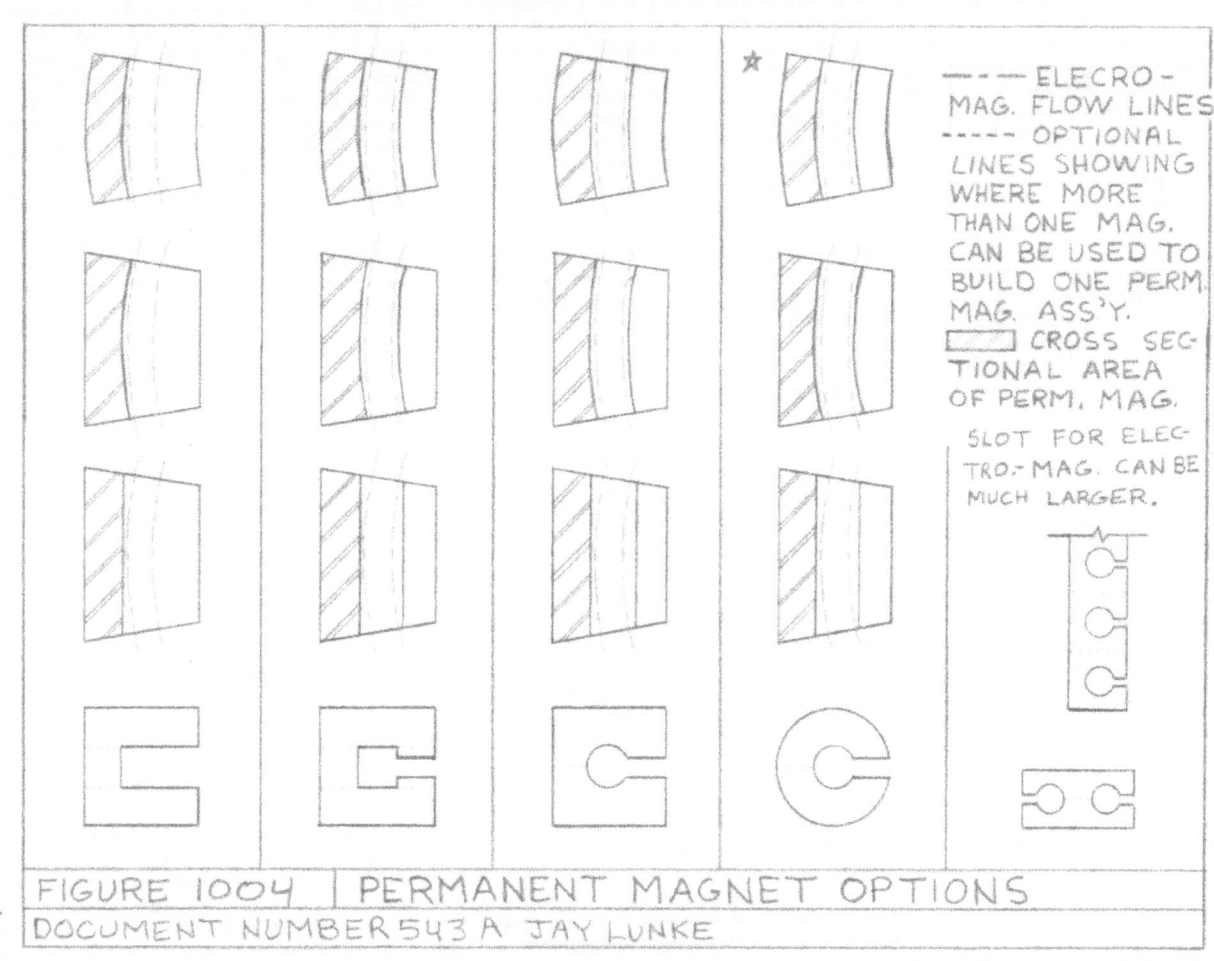

The permanent magnets are the most expensive parts on the motor. The ideal permanent magnet would be cylindrical in shape and made of the most powerful sustaining material you could find. To compromise for cost considerations, many other shapes can be used. For example, both the permanent and the electro-magnet magnet could be built into a squarer shape.

The major concerns when building the permanent magnet are as follows.

1. To use a material that has high permeability because of the physical magnetic rings repelling each other.
2. To use a material that will not deteriorate too much from the changing electro-magnetic field that travels through the channel of the permanent magnet ring.
3. Holes can be drilled into the permanent magnets to aid in mounting them, but this will affect the magnets performance.
4. The permanent magnet can be built up from one piece to many permanent magnets put together. The fewer pieces used to build the permanent magnet, the better it's performance will be.

Many methods can be used to mount the permanent magnets. The shape of the housing itself can hold the magnets into place.

The permanent magnets used in the flow through motor are more expensive to build than in most other types of electric motors. The many improved performance characteristics of the flow through motor makes it well worth the extra cost.

Electro-magnet assemblies;

Figure 1005 shows some examples of both the electro-magnets and mounting them onto the disc.

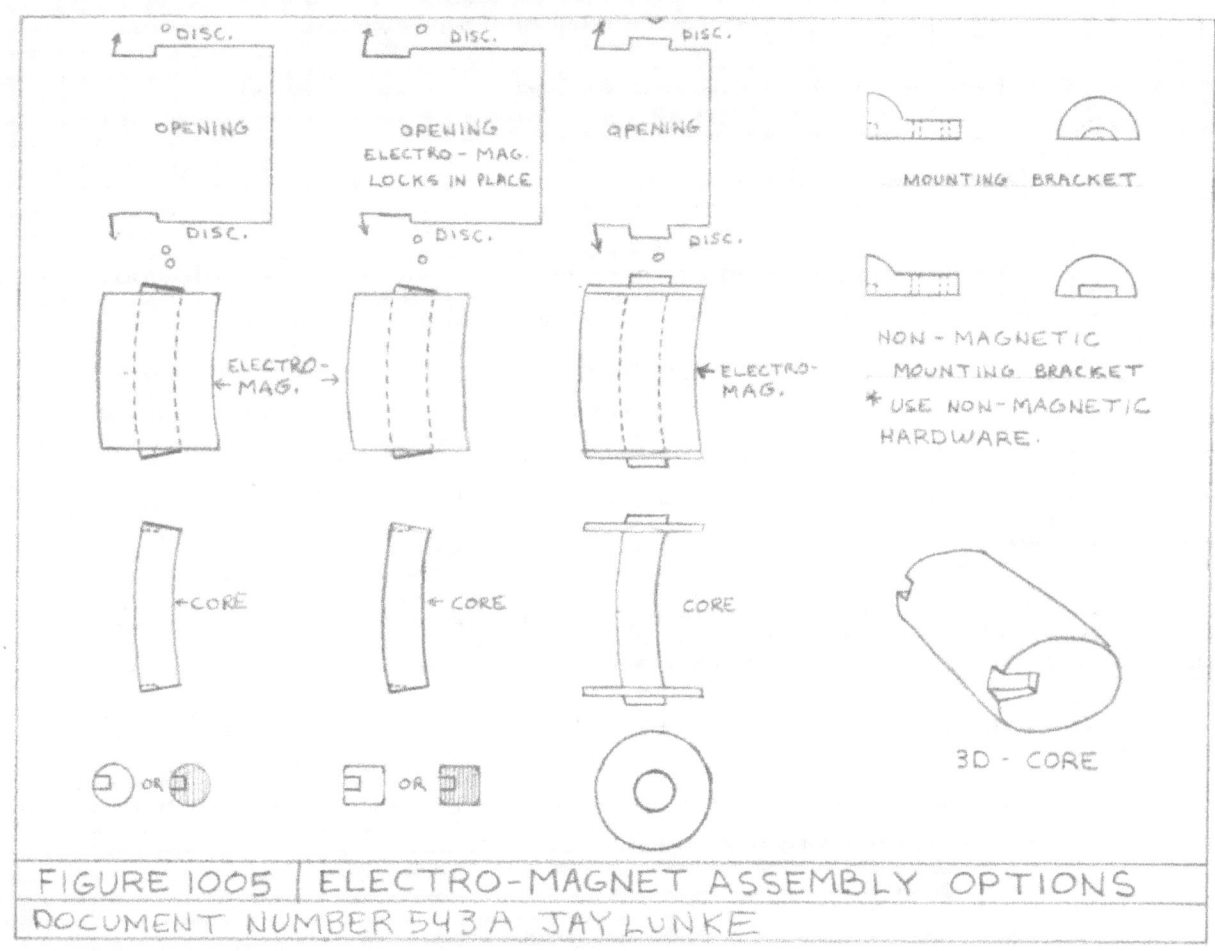

The core in the electro-magnet should be built out of a material that produces a high magnetic field.

This will provide a high flux density capacity for the electro-magnet. Also, special attention needs to be taken on the core materials coercive force. The core needs to do a good job in removing the residual magnetism in the core material.

The hardware and compounds used to mount the electro-magnet onto the disc need to be built out of non-magnetic materials. At the same time, they should be built with materials that will transfer the heat generated in the electro=magnet to the disc.

The electro-magnet will be built up with a wire wrapped around the core assembly. The most common wire to be used will be high temp. copper wire.

The wire gage used to build the electro-magnet has a direct relationship to the magnitude of the voltage used on the electro-magnet to bring it to its maximum flux density point. The smaller the gage wire that is used, the higher the voltage is required to bring it to its maximum flux density point. When the electro-magnet is travelling through the permanent ring, the voltage required to bring the electro-magnet to its maximum flux density point is affected. Also, the motors speed and loading have an effect on the electro-magnets maximum flux density point. These things need to be considered when designing the electro-magnet.

The electro-magnet can be designed to operate within many different voltage ranges. This magnifies the flow through motors variables and variety of applications.

The centerline of the flow through motor goes down the middle of the permanent magnet channel. The length of the electro-magnet and permanent magnet are the same at this line.

The number of electro-magnets mounted on the disc have a direct relationship to the motors torque.

The windings of the electro-magnet could be made of super-conductors. This would require a special cooling assembly to be installed. The cooling assembly will be discussed later.

The cylindrical shape is the most efficient electro-magnet design. In some motors the permanent magnets will be in a square shape in the channel area in order to reduce the overall cost of the motor. In order to provide the best operating characteristics, the electro-magnets need to travel as close to the permanent magnets as possible. When having permanent magnets with square channel shape, it is best to design the electro-magnets with a square shape as much as possible.

The core can be built up solid or be built up with laminations. Laminations provide better operating characteristics that reduce inefficiencies.

Since the most common electro-magnet disc assemble is built up with 50% electro-magnets through the channel area, 50% of this area is left for a variety of uses. Some of these uses could include electro-magnet mounting hardware, cooling coils, lubrication mechanisms, bearing assemblies, frequency and temperature sensor devices and other options are available

BEARING ASSEMBLIES:

The bearing assemblies hold the electro-mechanical assembly in its place in respect to the permanent magnet ring. Two basic types of bearing assemblies are used with the flow through motors. The mechanical bearing and the magnetic bearing assemblies, Figure 1008 shows some examples

examples of both types.

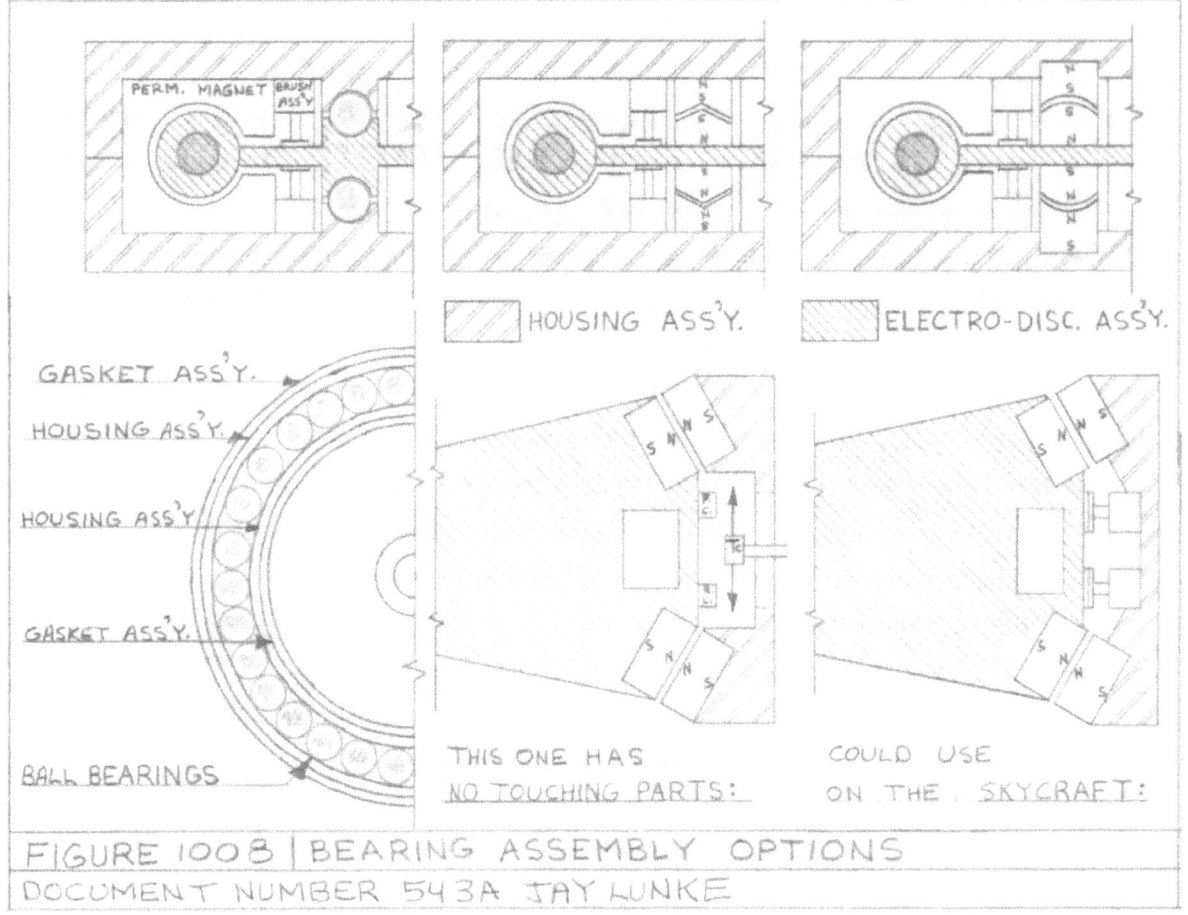

Both types can be mounted in a variety of locations and have their advantages and disadvantages. The mechanical bearing assemblies are much better at low speeds. The magnetic bearing assemblies on the other hand, are much better at very high speeds.

For the mechanical bearings, there are many parts on the market that can be used. In some applications, special bearing assemblies are required. These bearing assemblies can be separate or be a part of the housing and/or the electro-magnetic disc assembly. The mechanical bearing assemblies can keep much tighter tolerances between the permanent magnet ring and the electro-magnetic disc assembly. The mechanical bearing assemblies can handle heavy loads or pulls on the motor.

One of the disadvantages for the mechanical bearing assemblies is that the bearings will not hold up very long at super high speeds. At super high speeds, the bearing resistance is very high and the bearings will heat up or even freeze up.

The magnetic bearings do not have any touching parts. This means that the resistance is constant over a wide range of motor speeds. The magnetic bearing assembly will allow the motor to rotate at super high speeds. It also is a very low maintenance assembly. The biggest disadvantage of the magnetic bearing assembly, is the cost to build it.

The electro-magnetic disc assembly is very versatile and can have many configurations. The bearing assemblies come in a large variety to support many motor configurations.

MULTIPLE FLOW THROUGH ELECTRO-MECHANICAL MOTOR ASSEMBLIES;

The motor is not limited to one permanent magnet ring assembly and one electro-magnetic disc assembly. Figure 1012 shows an example of a multiple flow through electro-magnetic motor. One permanent magnet ring assembly can have from one to several channels in it to support as many electro-magnetic disc assemblies in the motor. The electro-magnets can operate independently or any number of them can be tied together. The independent electro-magnets disc assemblies can move at different speeds and/or direction from other disc assemblies using the same permanent magnet assembly.

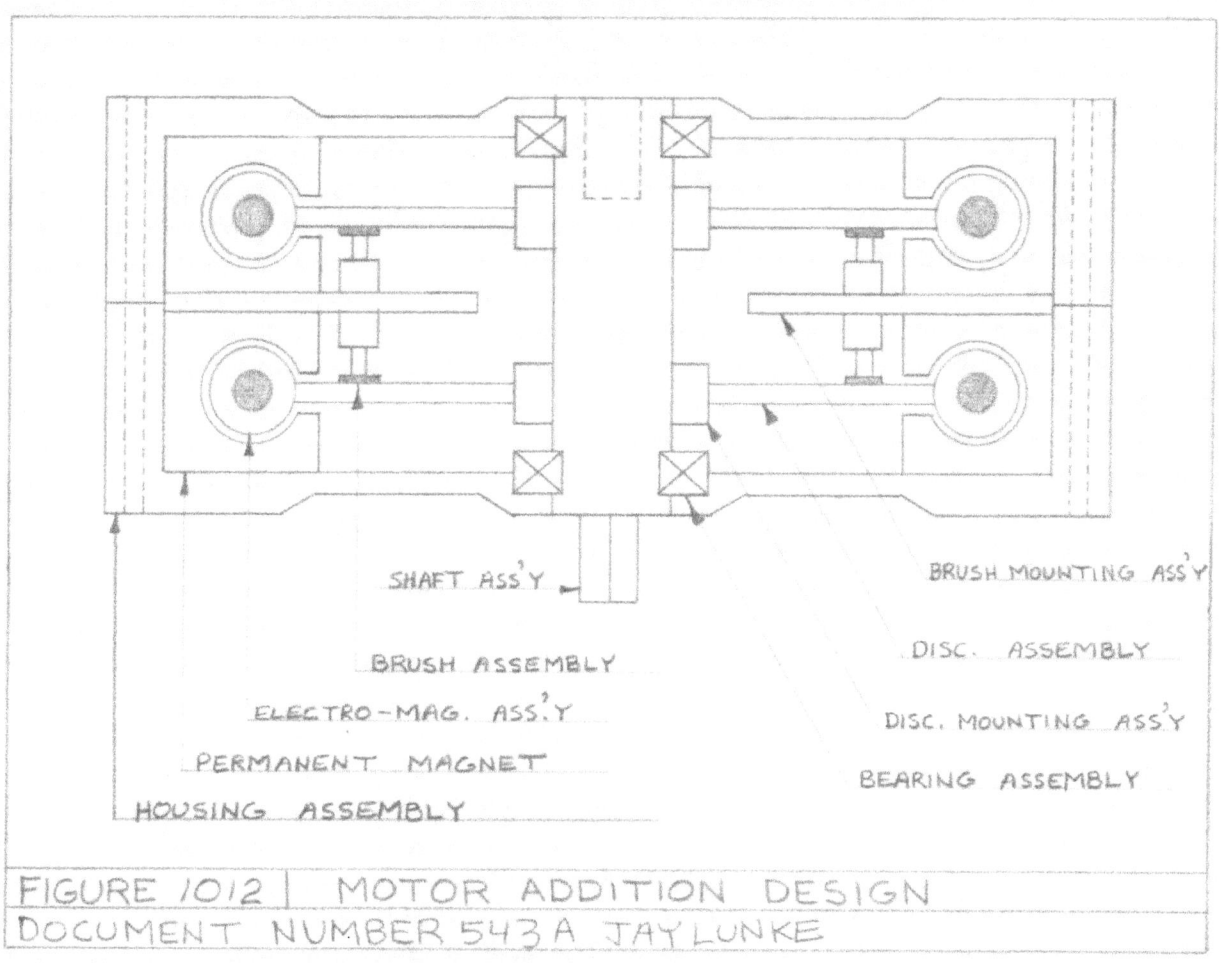

SHAFT ASS'Y

BRUSH ASSEMBLY

ELECTRO-MAG. ASS'Y

PERMANENT MAGNET

HOUSING ASSEMBLY

BRUSH MOUNTING ASS'Y

DISC. ASSEMBLY

DISC. MOUNTING ASS'Y

BEARING ASSEMBLY

FIGURE 1012 | MOTOR ADDITION DESIGN
DOCUMENT NUMBER 543A JAYLUNKE

One large advantage of a multiple electro-magnetic disc motor is that the disc's can be operated out of phase with each other to provide a more constant torque without startup problems on the total motor assembly. See 1023 for details.

The more electro-magnetic disc assemblies there are, the greater the motor power becomes. Also, the motor becomes more variable to additional applications.

The electro-magnetic disc's do not have to be very far out from the permanent magnet assembly before they can be attached to another assembly. They can be attached to many types of assemblies. One of these is a drum assembly shown in figure 1024. The drum could be part of a rock tumbler or cement mixer. It could be a drum in a processing plant, ext. Other attachments and applications will be discussed later.

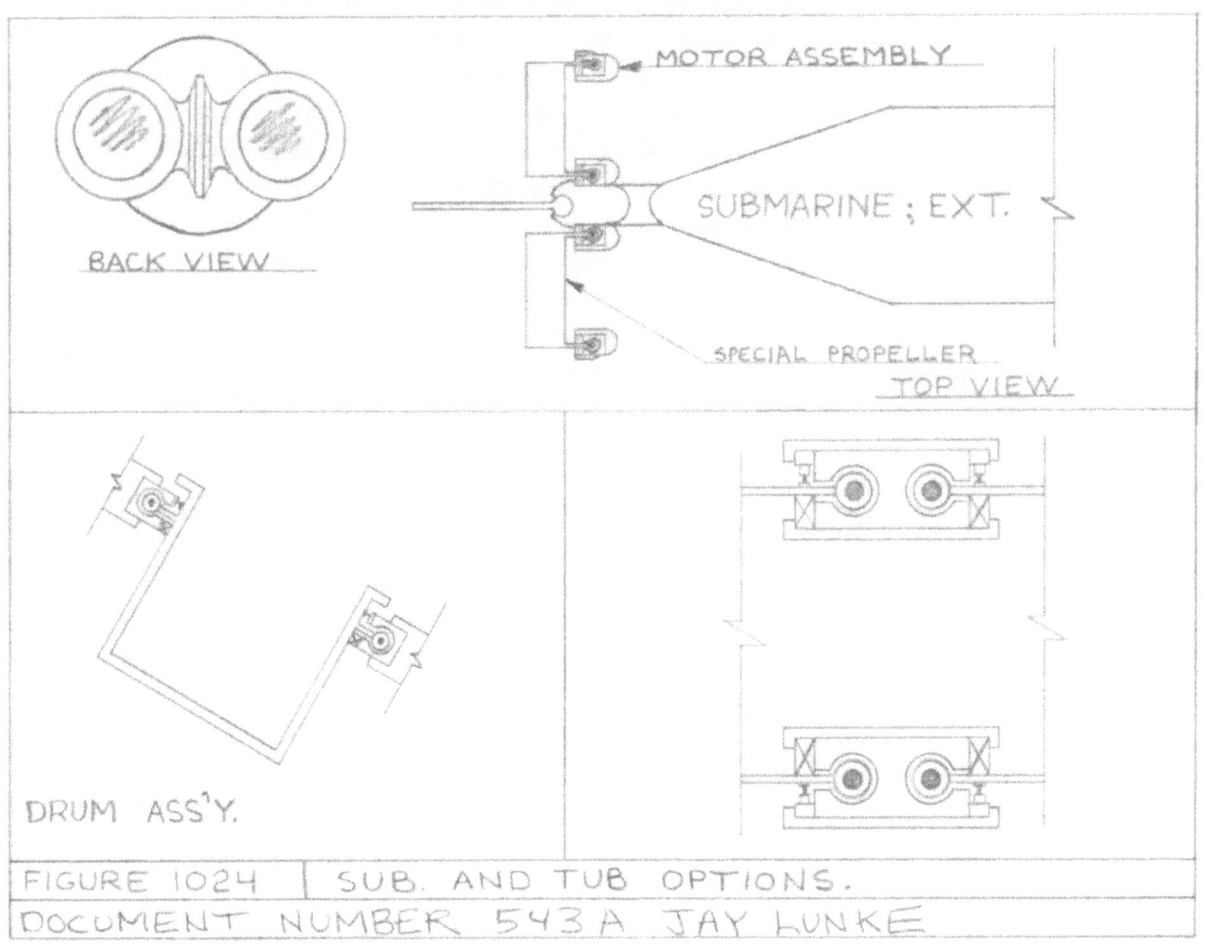

Flow Through Motor Angles

So far, I have only discussed the angle that the disc comes into the permanent magnet assembly as being in the inner circumference. The electro-magnetic assembly in fact can come out in any direction from the permanent magnet assembly. Not only it comes out in one direction, but in more than one direction having more than one electro-magnetic disc coming out. When more than one electro-magnetic disc assembly is present, they do not all have to come out at the same direction. The disc assemblies can be tied together or operate separately from each other. They can operate either at the same speed and direction, or at their independent speed and direction. See figures 1015, 1016, 1020, 1024 for some examples.

The direct side and outer circumference motors would not be connected to a shaft in most cases. The disc assemblies can be more closely attached to the application assembly to reduce the total assemblies required to get the job done. This will provide better efficiency for the many applications. The disc assembly could be tied directly to the wheel, saw blades, gear, paddles, pully, torque converter, fan blades, propeller, ext.

With in addition to motor angles, comes a large variety of motor styles with multiplied variety of applications. With less moving parts there is less maintenance and expense with these motors.

Motor Addition Design;

This is a concept of having a total package built up of motor module assemblies. The motor modules can easily be added or subtracted from the total motor package. As an application changes its power demand, it would be easy to compensate for it. The motor modules can be designed to easily attach themselves to another motor module. The shaft assembly would require a special design to be able to attach to and be attached to by another motor module. See figure 1012 for one example of a motor module assembly.

The number of motor modules mounted together to create the total motor package is called the motor string. For an example of how this would work, let us say that electric cars are sold with 4, 6, and 8 string motors. The motor you have in your car is a 4-string motor and you would like to have a car with a better power performance. Instead of selling your 4-string car and buying a 6-string car, you could easily have two motor modules installed to make your car a 6-string motor car. This makes the flow through motor more versatile in working with applications changing requirements.

POWER CIRCUITS;

The power circuits built up of all the parts used to bring electrical power from the power source to the electro-magnet. This includes all of the control circuitry. There are many types of power circuits. The power circuits range from a simple D.C. motor that is mechanically switched, to a computer-controlled variable resonate oscillator frequency and band width with load, temperature, and hysteresis compensation.

To start with, we will look at the connections to the electro-magnet. The electro-magnet had the two wires coming from it. Changing the direction of the current through the electro-magnet, changes the direction of the electro-magnetic field. You would switch the power wire and ground wire as the disc rotates. The one advantage to do this is that you could operate the motor on one power source. The disadvantage is that you would have to be switching two wires all the same time instead of one. It would also limit some of the power control options for the motor.

The best connections to use on the electro-magnet is to tie one side of the electro-magnet to ground and switch the other lead between a positive supply circuit and a negative supply circuit. Since the one lead is at ground potential at all times, it can be either connected through a brush assembly or tied to a conductive disc. assembly. The conductive path would go through the shaft for the ones that have one. The shaft would then be grounded. Grounding the disc assembly also provides safer operation at the applications level from electric shock.

Most motors have several electro-magnets in them. The electro-magnets can range from each electro-magnet having its own power circuit to all of them being tied together and operated with the same power circuits. One of the more common arrangements is to have 50% of them tied to one power circuit and the other 50% tied to a different power circuit. In this arrangement, even though the power circuitry is different at the electro-magnets, the signal conditioning circuitry can be the same. You would need to phase shift the signal before it goes to the other power circuit.

There are three major power signals to operate the electro-magnets. They are DC, AC, and a variable resonant tuned oscillator signal. The two major power controls on these signals are signal magnitude and frequency band with control.

No matter what power you have, you need to get it to the electro=magnet. If the electro-magnet assembly stays stationary and the permanent magnet ring assembly rotates, then you can have direct wire hook ups with the electronic switching. If the permanent magnet ring is stationary and the electro=magnet assembly rotated, then you need some kind of either brush assembly, pick-up coil or optical circuitry. These assemblies could provide either switching or constant contact for electronic switching. The constant contact is preferred because it has less noise, less wear and is very efficient.

The brush assembly can be mounted either on the housing or disc. Assembly. When the brush assembly is mounted on the housing, the contact assembly will be mounted on the disc assembly. If the brush assembly is mounted on the disc assembly the contact strip assembly will be mounted on the housing assembly. The contact strip assembly can be built for very long life. The brush assembly will last a long time but will need to be rebuilt or replaced before other motor parts. Replacement of the brush assemblies can be quickly done without major disassembly when designed into the motor. Figure 1006 shows some examples of brush assembly options.

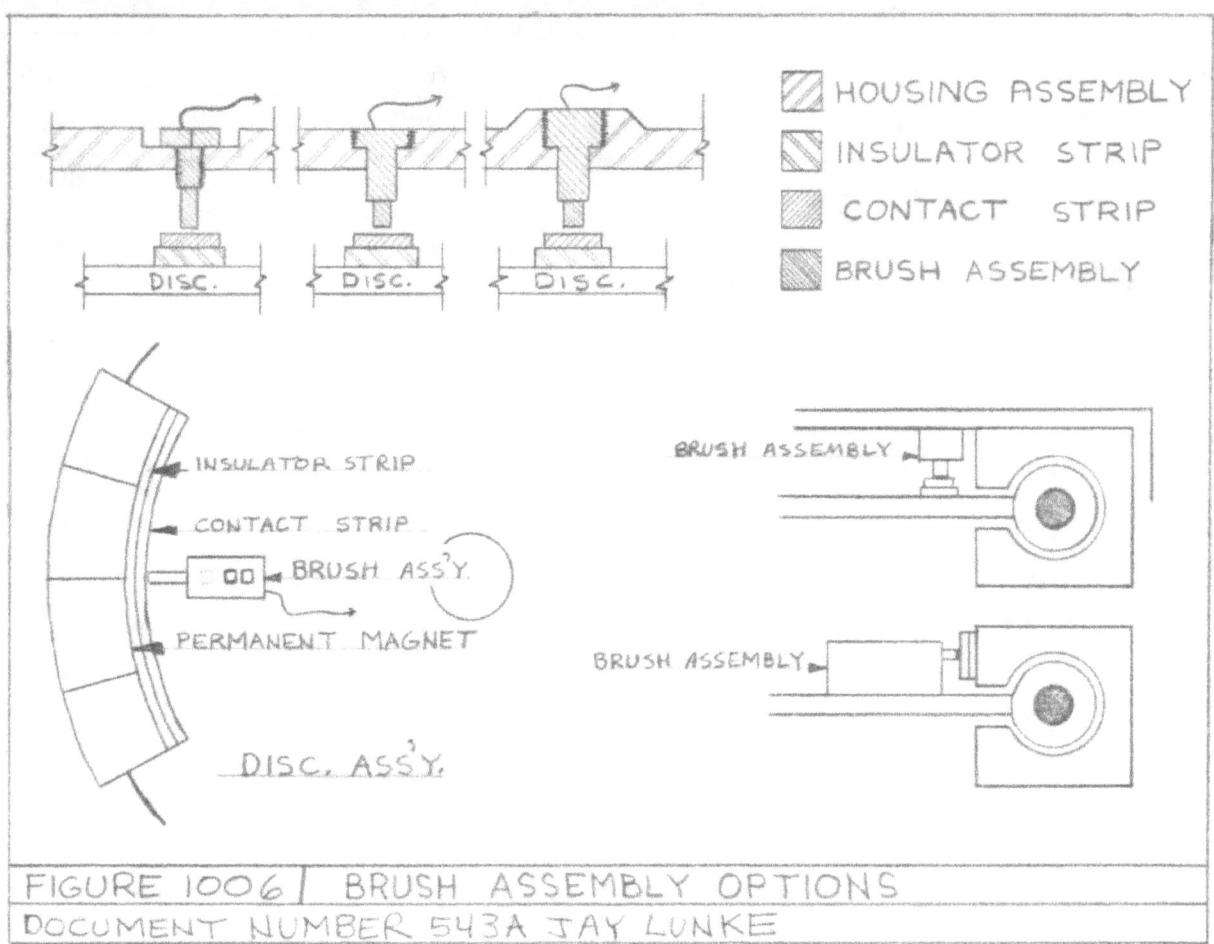

FIGURE 1006 | BRUSH ASSEMBLY OPTIONS
DOCUMENT NUMBER 543A JAY LUNKE

Now to look closer at the power signals that can operate the motor. The DC signal is a DC voltage that is switched either mechanically or electronically to the electro-magnets. The DC motor is the least efficient of the three types but it is simpler and cheaper to build. All three power signals can be controlled by signal magnitude and frequency band width control. Signal magnitude in the DC signal would be a variable DC voltage control. The bandwidth control in the DC is to control the percent of time the DC voltage is applied during the torque power cycle. By controlling the band width of power per cycle, you can adjust the bandwidth to the best part of the torque power curve. The middle of the torque power curve is the most efficient part of it. The squarish shaped torque power curve is very efficient itself; the bandwidth control adds to it.

The more power that is required for the motor, the wider the bandwidth would become. Controlling the bandwidth is more efficient to the motor than signal magnitude control. The bandwidth control is more expensive to build. You do have the option of combining both controls on a motor to provide an even better efficiency output over a large range of operating conditions.

Now to take a look at the AC motor signal. Many types of AC signals are available. The two most common of these are the sinewave and square wave. The square wave provides the most power to the motor while the sinewave is the most efficient of the two. The sinewave is more efficient because the majority of the power comes from it when the motor is in the most efficient part of the torque power curve.

The speed of the motor is determined by the operational frequency. The operational frequency in the bandwidth control circuit is not the same as the AC frequency. The smaller the bandwidth is, the greater the AC frequency is in comparison to the operational frequency. Like the DC signal, the AC signal can also be magnitude/bandwidth controlled. The bandwidth control circuitry can be done either mechanical or by electronical means. One option of the mechanical means is to have a taped contact strip assembly with the position of the brush assembly adjusted for the desired bandwidth. The electronic circuit option is more efficient and versatile for most motor options.

The most efficient and desirable power signal to use is the variable tuned resonant oscillator circuit. Figure 1009 shows one of these circuits in its basic form. The theory of operation is as follows. A coil in this case it will be the electro-magnets, and a capacitor connected in parallel with each other, have a resonant frequency. It is at this point that the motor is operating at its most efficient point. In order to operate the flow through motor at its resonant frequency through different operating conditions, you need two variable capacitor circuits. You also need two switches to operate with each capacitor. One to connect the capacitor to the power source to recharge it back to its full potential. The switches that are used, can be either mechanical or electrical switches. The electrical switches are more efficient and are open to other motor operations.

Note: The following Figure was modified due to an error in the original drawing. The Magnetic polarity of the electro-magnet is not shown in the circuit. The direction of the flow of electrons through the electro-magnet determine the polarity of the magnetism in it. While the electro-magnet is operating in the tank circuit, the torque and magnetic polarity go though one complete cycle when the tank circuit is connected with the capacitor. In order to allow the capacitor to recharge back to the full potential, two capacitors are used in the circuit in order to alternate being cycled through both the charging time and the tank operation time.

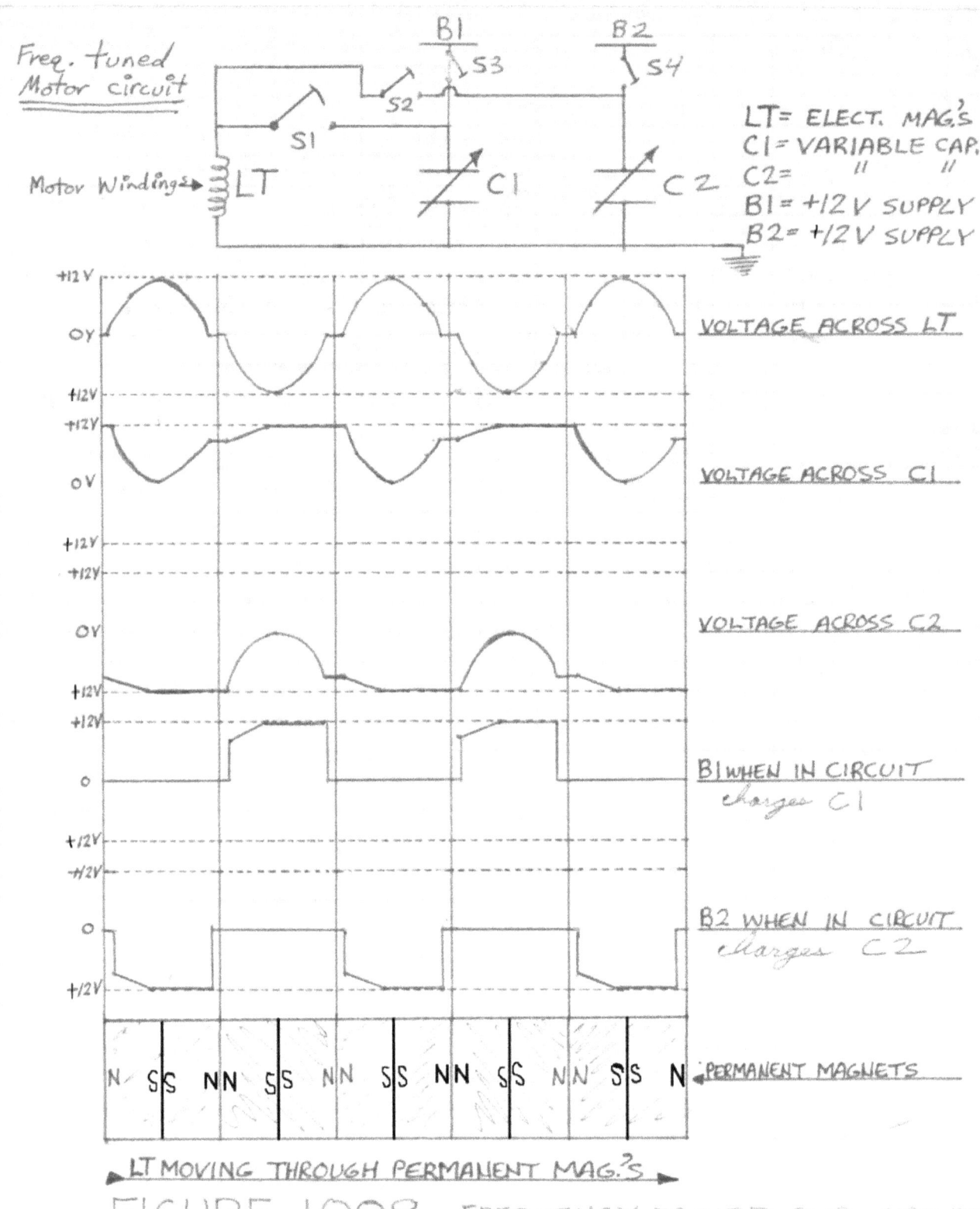

Freq. tuned Motor circuit

Motor Windings

LT = ELECT. MAG's
C1 = VARIABLE CAP.
C2 = " "
B1 = +12V SUPPLY
B2 = +12V SUPPLY

VOLTAGE ACROSS LT

VOLTAGE ACROSS C1

VOLTAGE ACROSS C2

B1 WHEN IN CIRCUIT
charges C1

B2 WHEN IN CIRCUIT
charges C2

PERMANENT MAGNETS

LT MOVING THROUGH PERMANENT MAG.'s

FIGURE 1009 FREQUENCY TUNED PWR. CIRCUIT

DOCUMENT NUMBER 543A JAY LUNKE

So, again, when you have a coil and a capacitor tied together, they will have a resonant frequency where the current flows back and forth between them. These circuits are found in all the old tube TV sets and many other circuits. The current flow is very efficient moving in this type of a circuit. The circuit has only wire resistive and some induction losses in them. When the current flows back and forth in the coil, a magnetic field is created. The polarity of the magnetic field changes direction as the current flow changes direction in it. The electro-magnets would replace the coil in the circuit. The back EMF occurs when the coil is releasing its energy back into the capacitor. This back EMF is in the opposite polarity as when the power is going into the coil. From the time the current flows into the electro-magnet to its fullest potential, the electro-magnets have moved through one complete torque cycle. When the back EMF starts, the physical position of the rotor has moved into a location that needs this reversed polarity to create positive torque on the motor assembly. It is during this time that the electro-magnets move through another torque cycle. Also, during this same time, the capacitor then captures the energy minus the resistive and induction losses back into the capacitor. The switching is made to then charge the capacitor back to its full potential.

The reason I have two variable capacitors instead of one is so that the capacitors can be topped off to a full charge before they are used again. When capacitor one is disconnected from the electro-magnet then capacitor two is connected to the electro-magnet. The process of current and magnetic force repeats itself the same way with the new capacitor connected to the circuit. Note that when switches to the capacitor is open, the capacitor locks in the power it captured and holds it for as long as you want to hold it. The capacitor that is not connected to the electro-magnet is connected to the power supply. The reason I have two capacitors in the tank circuit is because I need to allow time for the capacitor that was just disconnected from the electro-magnet and connected to the power supply time to top off the charge it has to full potential. It would be nice if this circuit should operate almost as efficiently as the TV tank circuit. The losses will be more in the motor applications. For each operating condition of the motor, the capacitor value is adjusted to the resonant point. This point is the most efficient energy usage point of the motor assembly.

The reason the resonant circuit is so efficient is because more than some of the energy used to create the electro-magnetic field is save back up in the capacitor. When the power supplies recharge the capacitors back to full potential, it does not require a lot of energy. The capacitors need to change values with changing motor operating characteristics. Figure 1010 shows an example of an electronic variable capacitor circuit assembly.

One thing to note here is that the object is to recover most of the electrical energy as possible that was used by the motor to move through two torque cycles so that you can use that electrical energy again. In the electrical circuit applications using the tank circuit, the resonant point has to be rite on the optimal resonant point. But on the motor power circuits it does not have to always run at the optimal resonant point. As long as we are capturing some of the power and reusing it again, the efficiencies of this circuit in conjunction with the motor design will be great.

With the fast computer circuitry that are on the market place today, if a snapshot of the time it takes for the electro-magnet to reach its peak voltage just before the current starts to flow in the other direction was taken, then that information could be used to adjust the capacitance value to keep the motor at its optimal operating point. With this type of information to the control circuit, none of the other compensation information would be needed. This is because all of the other information is being used to anticipate this snapshot that was just taken by the control circuitry.

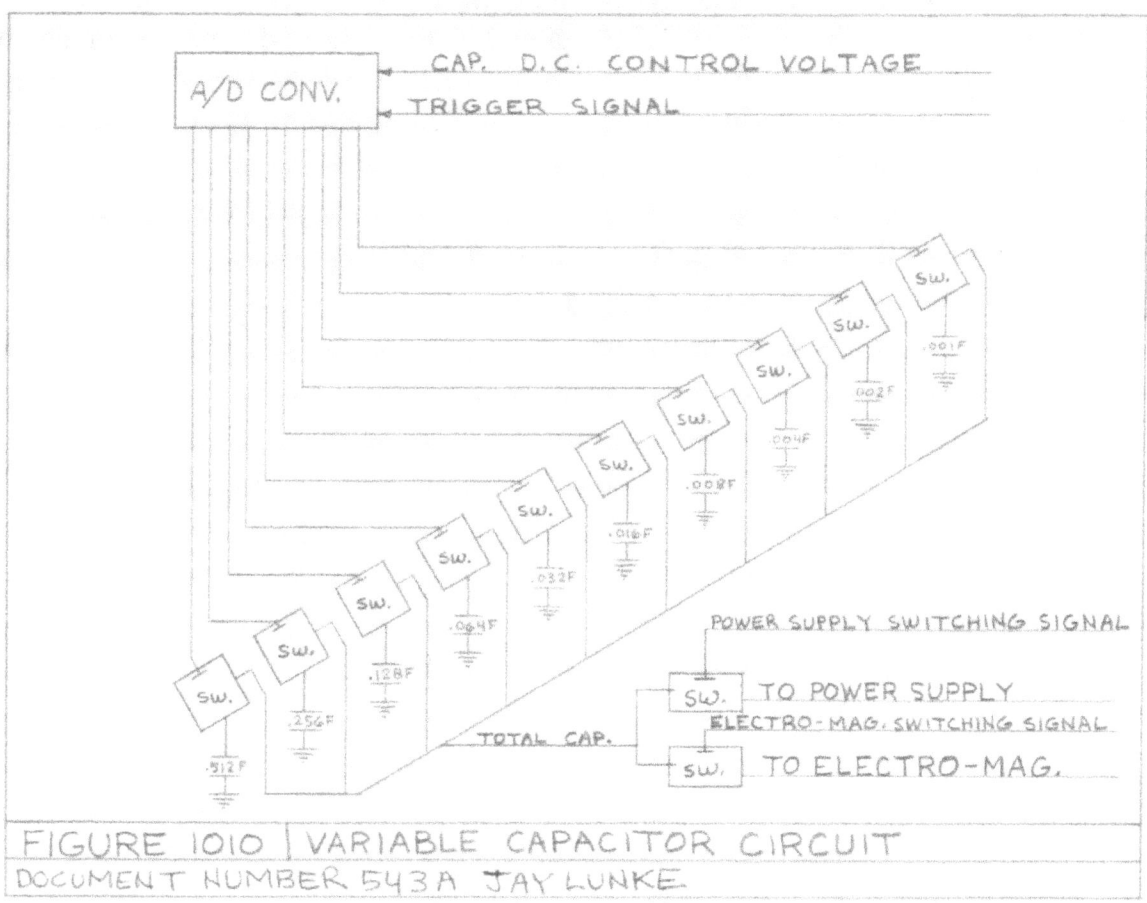

FIGURE 1010 | VARIABLE CAPACITOR CIRCUIT
DOCUMENT NUMBER 543A JAY LUNKE

The resonate oscillator signal circuit can work in conjunction with band width control circuitry. The switching algorithms would need to change in the tank circuitry for this to happen. This would make a very efficient motor operating over a wide range of operating conditions. Magnitude control could also be added to this scheme.

The magnitude and bandwidth controls pertain to the electro-magnet control. Since some of the motors have many electro-magnets, the operating performance can be controlled by controlling the number of electro-magnets that are connected to the power circuitry at any one time. An economical way to do this is to have whole electro-magnetic disc assemblies switched in and out from the power circuitry. Since individual electro-magnets would operate in the more efficient portion of their torque power curves, the overall motor efficiency is raised with this type of motor control in conjunction with the other ones.

Control circuits could include circuits to compensate for such things as the delay time between the electrical signal and the electrical-magnetic field. Core material hysteresis, varying load conditions, motor speed, and temperature changes.

Figure 1011 shows a block diagram of circuitry that could be used to control a vehicle.

In using control circuitry, it is also important to pick up different signals that indicate the motors performance. Some of these signals could be motor speed, temperature, power consumption, and even parts of the power cycle to analyze for the resonant oscillator control circuitry. A control circuit making adjustments off from signal samples could save time and money. Instead of compensating for each individual variable separately, it would compensate for all of the variables by compensating off of the signal samples in real time during the motor operation.

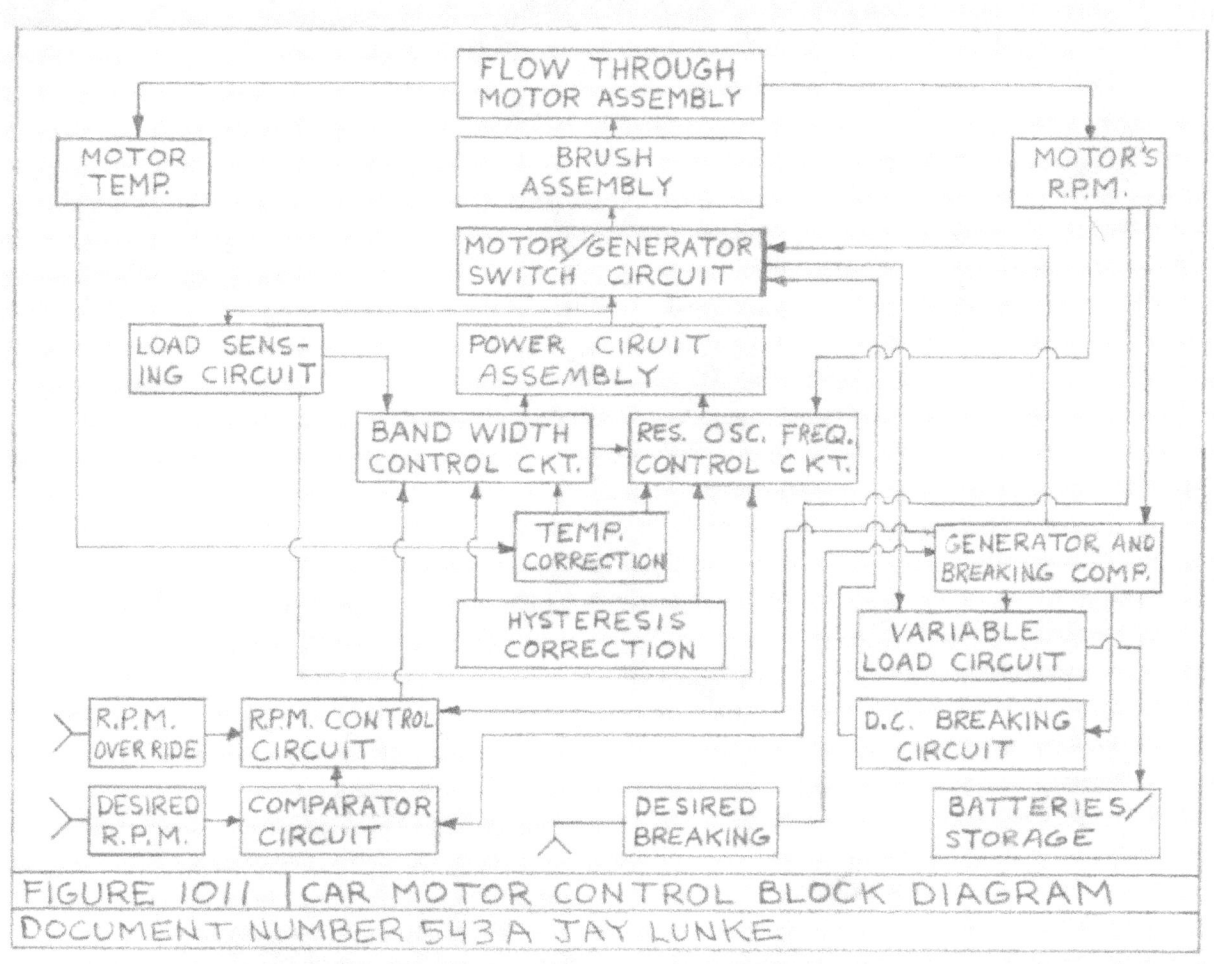

Many of the power circuits would run off of batteries. They could be mounted in such a way, to provide for a quick connect/dis-connect assembly. Then recharged batteries could quickly replace the discharged ones.

The flow through motor provides a very wide range of power circuit control compared to armature motors. The motor circuits provide some super high efficiency ratings compared to other motors.

The control circuitry has about as many advantages as the flow through motor itself. The combination of making the best out of both, adds up to a very powerful efficient motor package.

Cooling Coils and Fin Assemblies;

Cooling coils can be used when electro-magnets are made up of super conductors. The cooling coils could be mounted next to the electro-magnet assemblies. Both the coils and mounting hardware would be built with non-magnetic materials. They would be built with materials that have good thermo-conductivity. Figure 1013 shows an example of a coil and fitting assemblies. If the electro-magnet disc assembly rotates, then you need a special fitting assembly to move the cool fluid into and out of a moving part. If the permanent magnets are part of the moving assembly, then the plumbing becomes much easier because it does not have to travel through any moving parts. The flow-through motor using super conductors can only mean super performance all around.

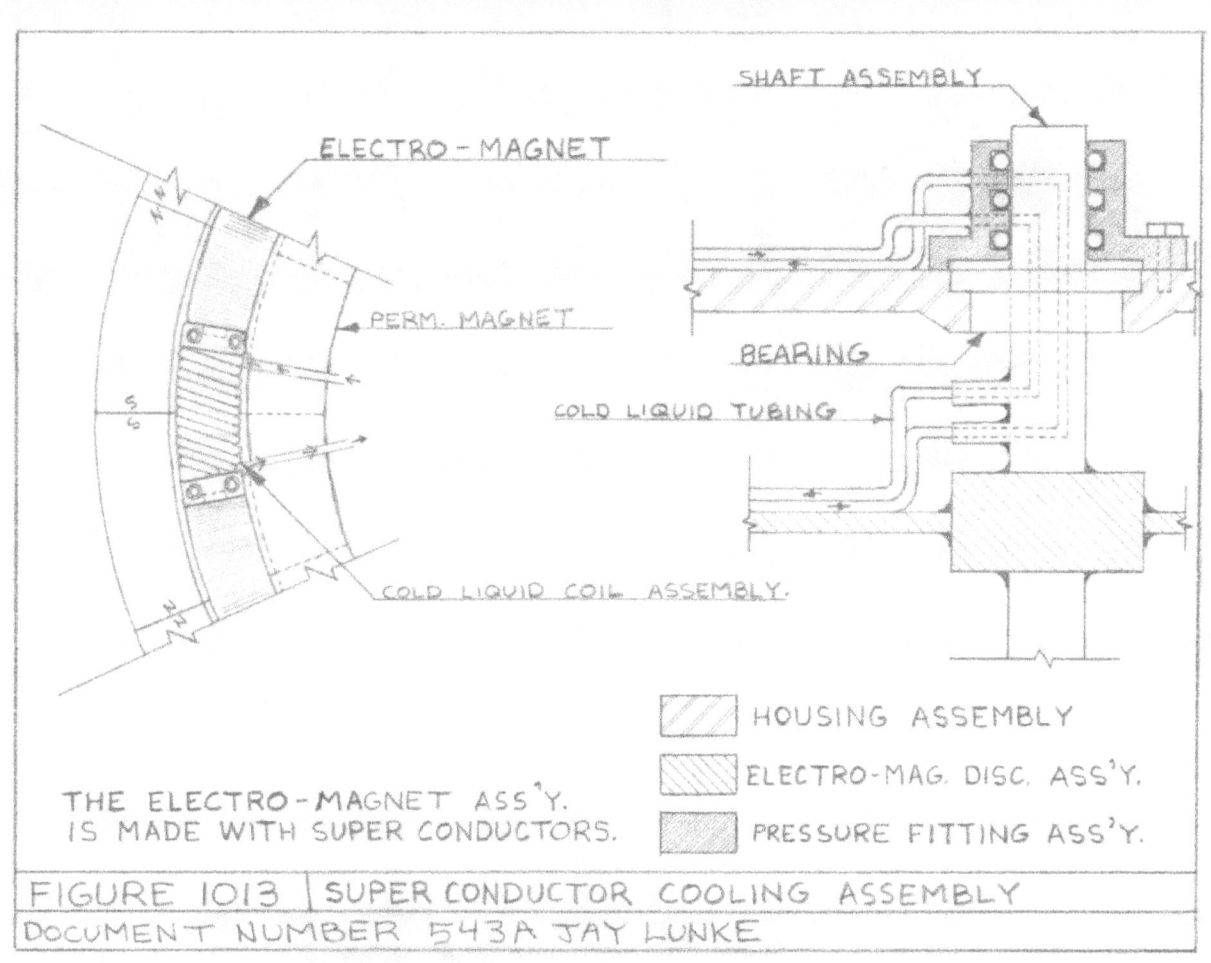

The majority of the flow through motors use copper wire. With some applications, it will be necessary to cool off the electro-magnets. The electro-magnets can be thermo-conductivity connected to the disc assembly. The disc assemblies can have fin assembly mounted on or into them the heat will transfer from the fin assembly to the air the fin assembly is moving through the motor. Figure 1007 shows some examples of these fin assemblies.

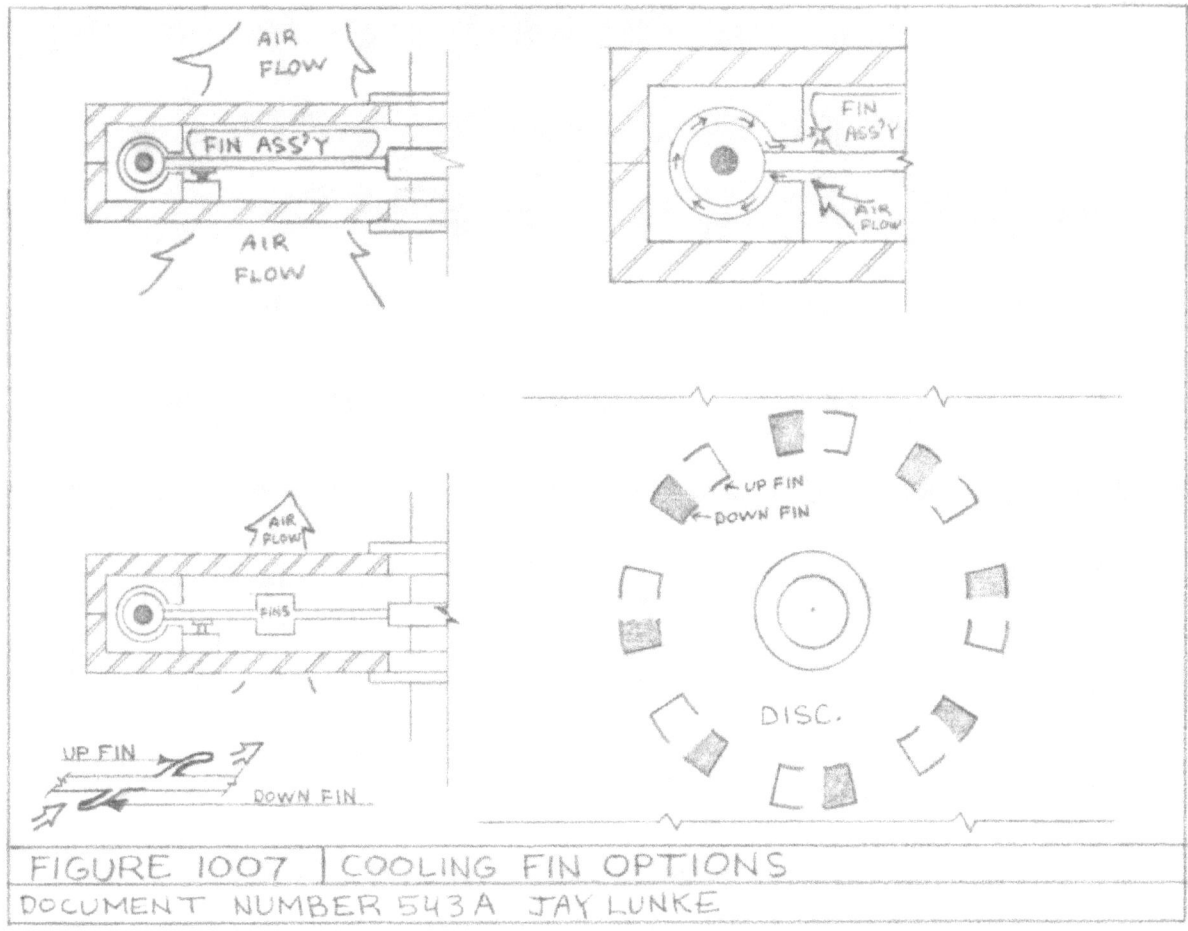

FIGURE 1007 | COOLING FIN OPTIONS
DOCUMENT NUMBER 543A JAY LUNKE

It is much easier to remove the heat from the flow-through motor than it is from the armature motor. The electro-magnets are spread out more in the electro-magnetic motor assembly. In the armature motor, all of the heat is produced in one area.

Vehicle Applications:

The flow through motors can be used on a variety of different vehicle applications. The motors with a shaft in the configuration can be hooked up like conventional motors and perform the same job more efficiently.

The flow through motors come in many other applications. The motors with a shaft in the configuration can be hooked up like conventional motors and perform the same job more efficiently.

The flow-through motors come in many other configurations that can be hooked up directly to the application assemblies. The configurations can be built into two modes. In one mode the permanent magnet ring would be stationary. The other one would be a stationary electro-magnet disc. assembly. This means you can have the work performed off from either one of these two assemblies. Figure 1015, 1016, 1020 1024, 1014, and 1000 show examples of the flow through configurations that can be used on land vehicles. Some of the configurations can be directly hooked up to wheels or track assemblies. Other configurations can hook up directly to a chain, belt, torque converter, ext.

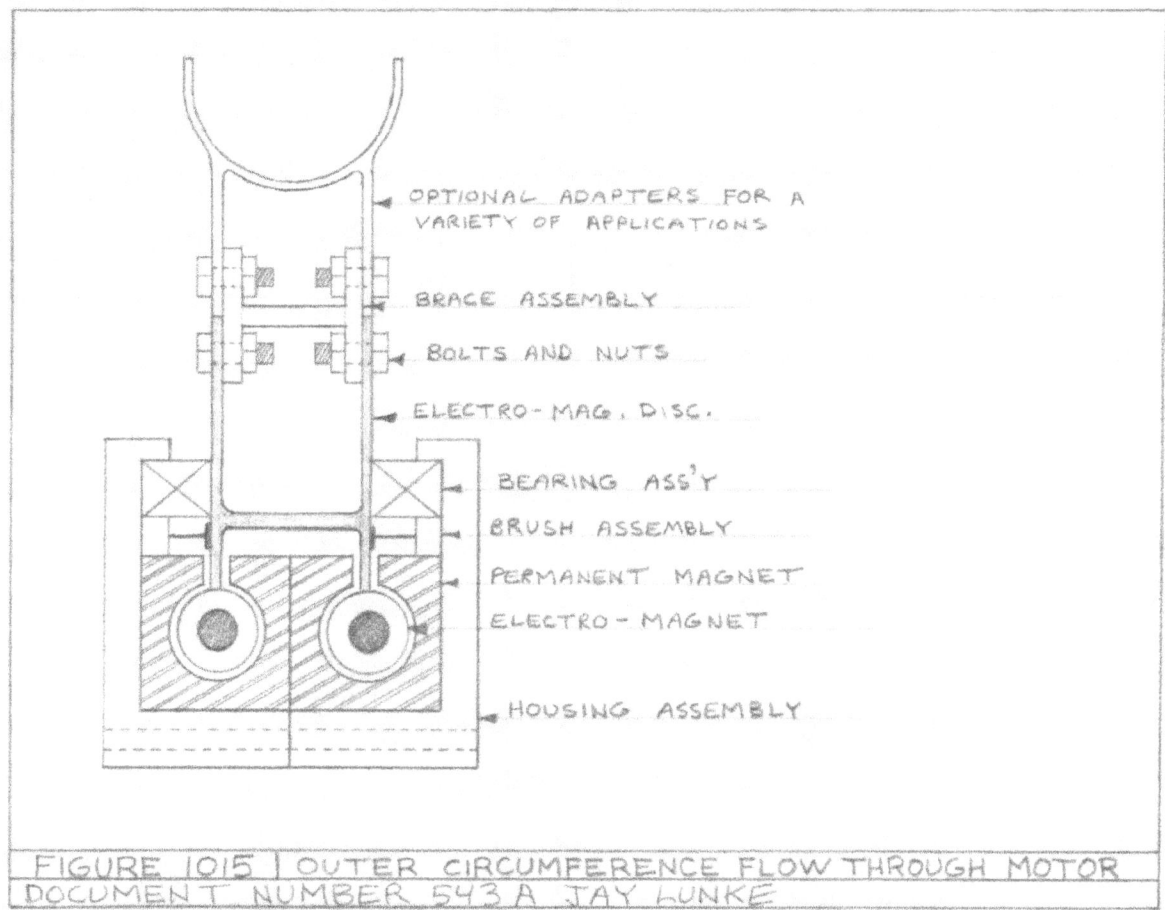

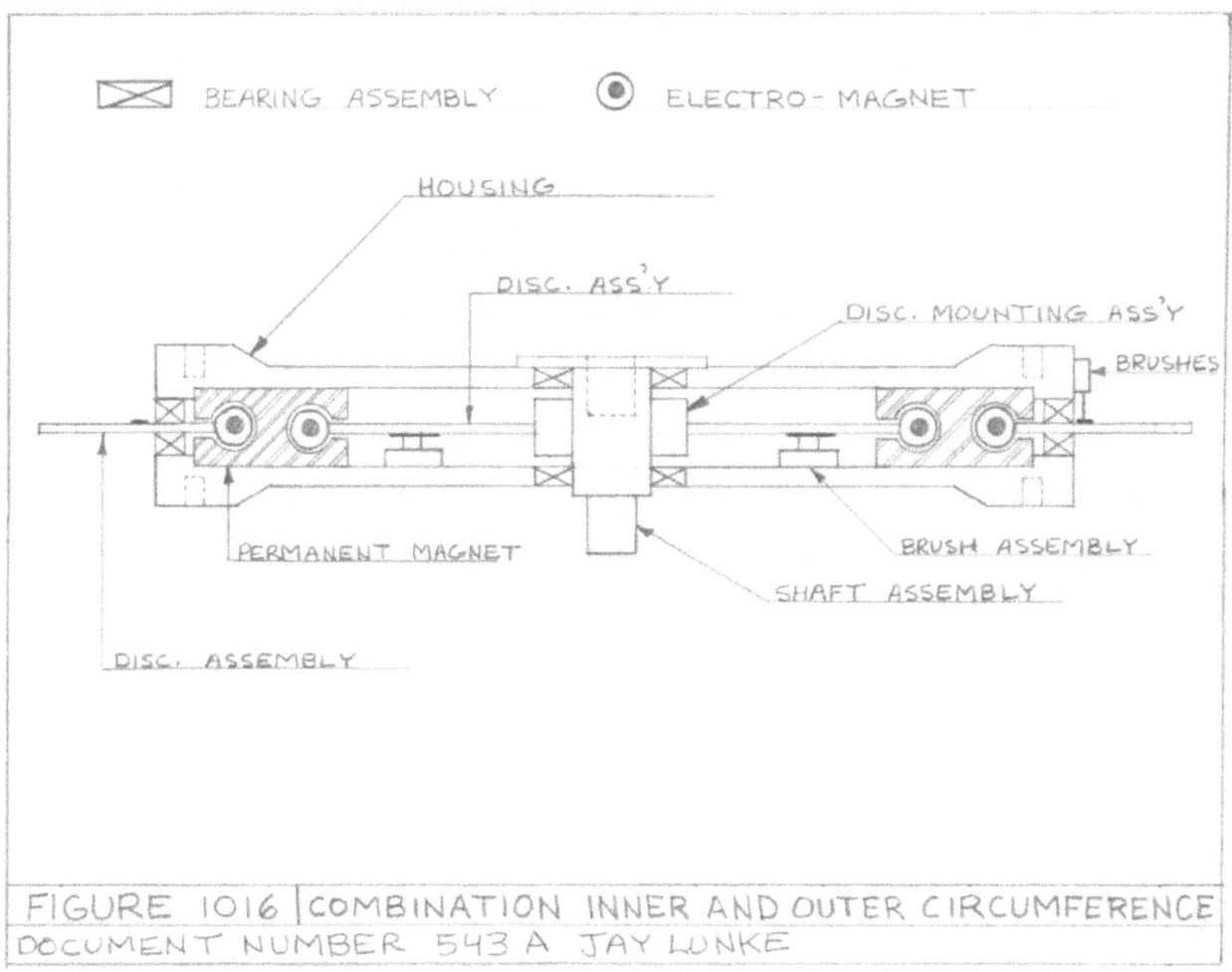

FIGURE 1016 | COMBINATION INNER AND OUTER CIRCUMFERENCE
DOCUMENT NUMBER 543 A JAY LUNKE

You could hook up one motor for each wheel to provide the vehicle with a lot more control and traction.

Each motor can change its direction. This means you do not need a transmission. The motors are very high torque and can operate at high speeds, so no gears are required for their motor vehicle applications. Figure 1021 shows some examples of vehicle transportation with a type of flow through motors not restricted to an electro-magnetic disc assembly and permanent ring assembly.

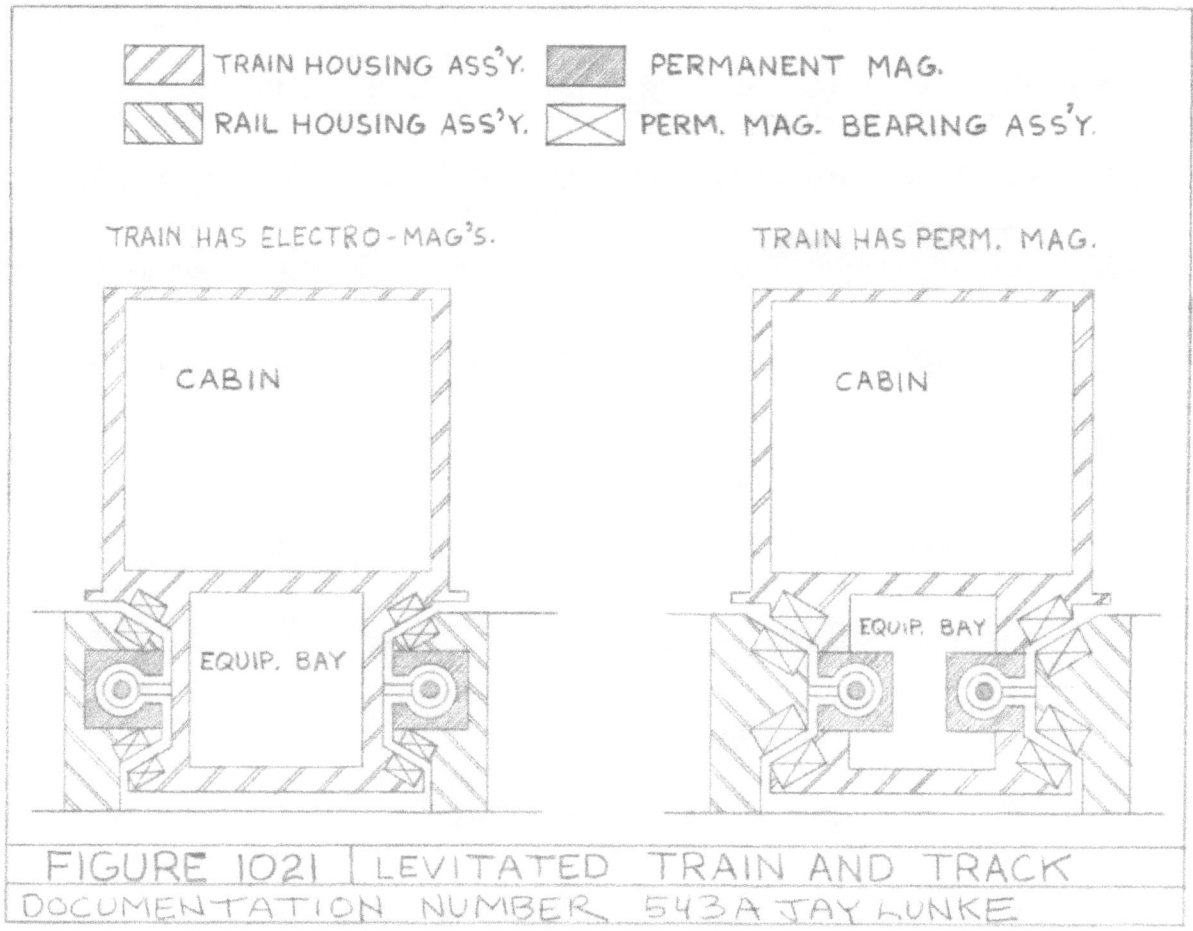

Here both the electro-magnets and the permanent magnets operate in a straight line. The vehicle can hold either the electro-magnets or the permanent magnets. The vehicles can stay on the ground or be levitated as shown. The levitated vehicles are capable of very high speeds. Using this type of vehicle, you are limited to only wide turns even with shorter cabs with motor assemblies mounted in the front and back portions of the vehicle. With the vehicle traveling at very high speeds, you would not want to turn too sharp anyway. These vehicles can be linked together into a train. When designing a levitated train with the flow through motor, you would want to use magnetic bearing assemblies to hold the vehicle in place with respect to the tack. The vehicle could be totally isolated from having any touching parts.

Flying "SKY CRAFT"

The motor can be made to any circumference you want. In the sky craft applications, the motors have a large circumference. The electro-magnetic disc assembly can be part of / or connected to a fan or propeller assembly. The basic sky craft is show in figure 1017.

The sky craft has two propeller assemblies. One moving in the clockwise direction and the other in the counter-clockwise direction. Both propeller assemblies are designed to move a large air flow from above the sky craft down through the propeller assemblies and out through the bottom of the sky craft. There are fin assemblies mounted in the sky craft to direct the air flow for desired maneuvering of the sky craft. The propeller assemblies have a large variety of flow through motor options that can be used to operate the sky craft. An example of the propeller assembly options is shown in Figure 1018.

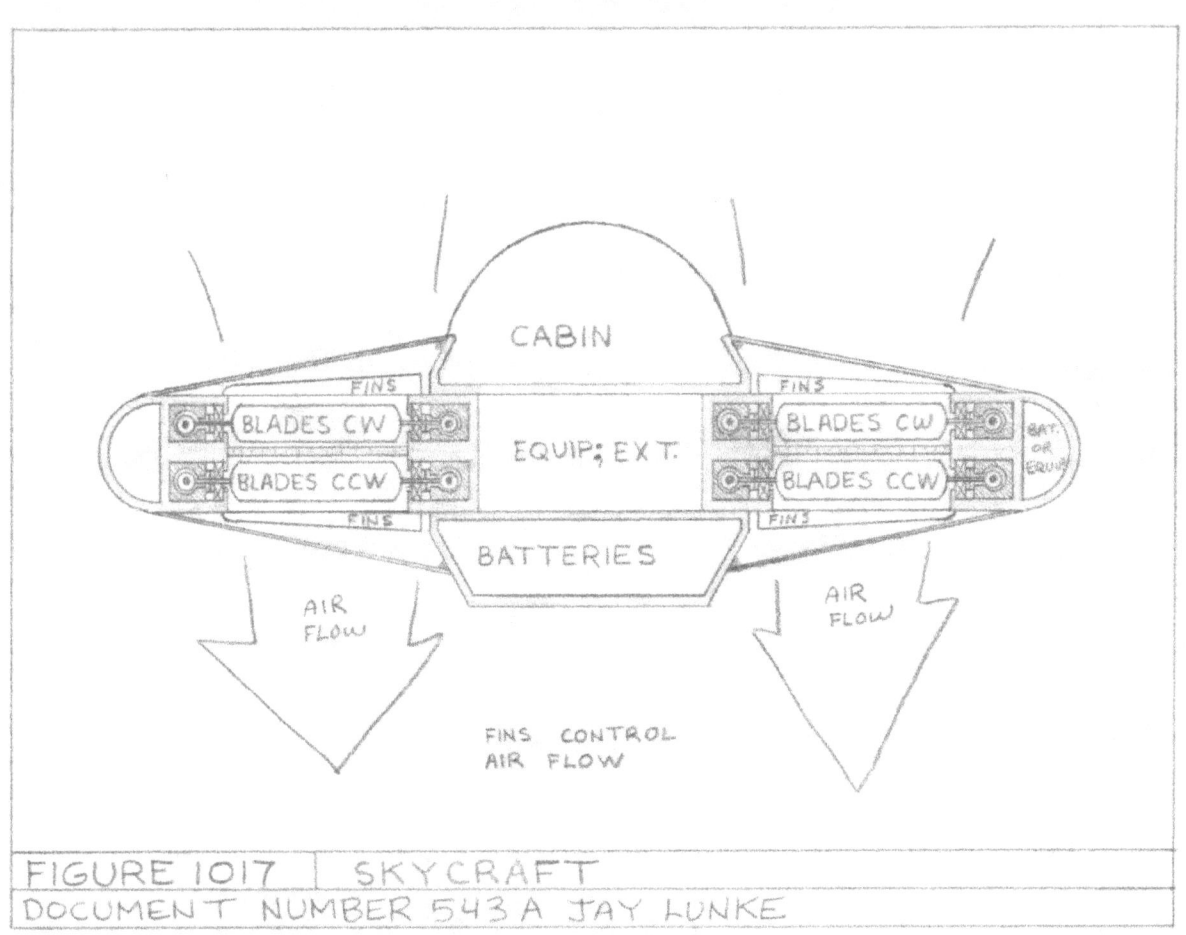

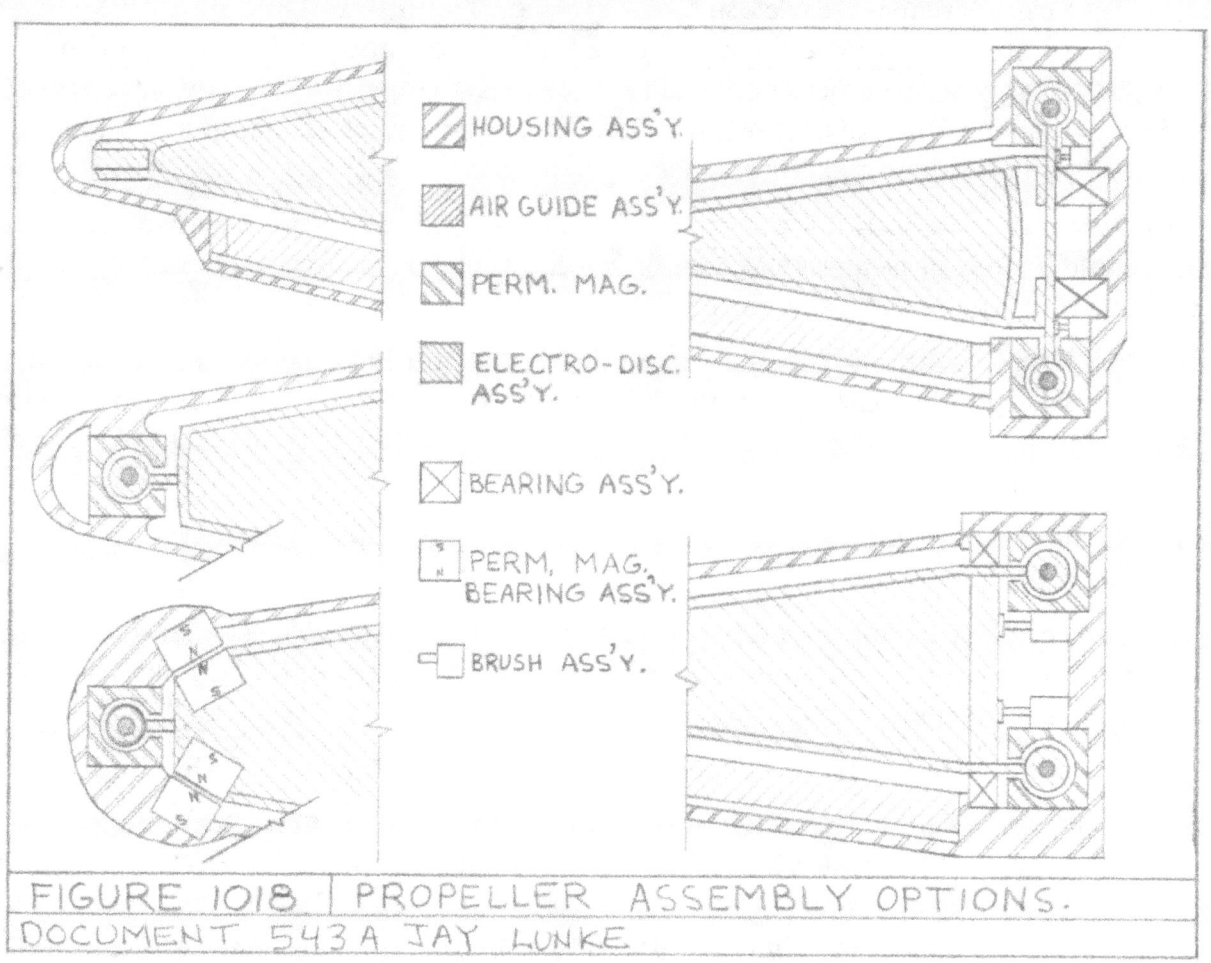

FIGURE 1018 | PROPELLER ASSEMBLY OPTIONS.
DOCUMENT 543A JAY LUNKE

In the sky craft with two propeller assemblies, the relationship between the speed of the assemblies, will control the rotation of the sky craft to change the direction it is facing. In a single propeller assembly sky craft, the craft will have the tendency to spin. To prevent this from happening, part of the fin assemblies' function is to direct the flow of air in a direction to compensate for it.

The part of the craft above and below the propeller assemblies are part of the housing assembly whose function is for providing a bracing network to hold the outer housing with the magnets in it in place. The bracing network can be over-laid by a screening network to prevent objects from flying into the propeller assembly and causing damage.

The weight of the electro-magnetic propeller assembly turning at very high speeds, gives the sky craft the characteristic of a gyro. It is this characteristic that will keep the sky craft very stable during its flight. At the same time the sky craft is very stable, it is very maneuverable. The sky craft can move in any direction through the air by controlling the motor speed and fin adjustments.

The sky crafts that use magnetic bearing assemblies instead of the mechanical bearing assemblies, can produce very high speeds and have low maintenance requirements. Also, the heat generated in the bearings is much less.

The sky craft requires a small area to land and take off in. The sky craft can have a redundant power and control circuitry to provide a backup system for the sky craft's safety.

The sky craft can maneuver very close to the ground. The sky craft could incorporate a quick connect battery system so that the sky craft could have a short down time before it can be flown again. The batteries could be rented or exchanged on an exchange system. The batteries could be mounted on an assembly toward the bottom of the sky craft for quick exchange time. The batteries could be broken into two battery sets for a backup system.

The inside part of the sky craft is the cab area. It can be used for a large variety of uses. Some of these uses could include passengers, cargo, military equipment, office, cabin, home. A group of sky crafts tied together could form a sky train.

The sky craft looks nothing like an airplane or a helicopter but the sky craft acts a lot like a helicopter. In fact, the flow through motors could be used to rotate the blades on a helicopter. The sky craft protects the blades a lot better than the helicopter does. The sky craft is almost maintenance free compared to the helicopter.

The larger the area the propellers cover in comparison to the cab on the sky craft, is greater the speed and control will be.

The sky craft can be used for private, commercial, or other usages. The sky craft would be very safe to operate and easy on the environment. In mass production, the cost of the sky craft would be close to the same as the helicopter.

Some of the options for the sky craft could include attachable pontoons for landing on water.

The sky craft can be built with as little as one moving assembly. One way to do this would be to install the control and power assemblies onto the electro-magnet propeller assembly. The control information can be electro-magnetically transferred to a pick-up coil off the control circuitry on the propeller assembly as shown in figure 1008. Optical switching control is another option.

The best way to provide the least amount of contact from the propeller assembly to the housing assembly, is to switch the major components around on the motor assembly. The permanent magnet ring can be mounted onto the propeller assembly. The electro-magnetic disc assembly can be mounted onto the housing assembly. A sensing coil can be mounted onto the housing to measure the relationship of the permanent magnets to the electro-magnets and feed it to the control circuitry. This would keep control and power assemblies off of the propeller assemblies. Eliminate the need for brush assemblies. A permanent magnet propeller assembly would have less maintenance, fewer parts, run more efficiently and have better performance characteristics.

The sky craft is very versatile, maneuverable, dependable and safe. It is an alternative for today's air travel.

Another option is to build replacements for the jet engines that use this new technology by adding several layers of motors either at different distances from each other or different diameters from each other to achieve increasing the thrust going through the device. These simulated jet engine designs would be able to be used on vehicles that operate under water. They could be added to two sides of the sky craft to increase the speed of air travel for the craft.

COMBINATION SKY
AND SEA CRAFT;

With some simple modifications to the sky craft, it can fly in the air, float on the water, and go under the water and maneuver around. Special gaskets can be installed to protest sensitive components from exposure to the undesirable elements under water. These gasket assemblies can be used to protect many other flows through motor configurations.

Another option for some motor configurations, is to apply a protective coating over sensitive parts, another option is to build the parts waterproof that come into contact with the environment.

This craft would need a cabin control system installed that would control the cabin pressure in relationship to the outside pressure. Without motor movement, the craft should be light enough to float on the water.

Like any other flow through motor configuration, when the power sources are switched, the propeller assembly changes direction in which it turns. It is at this point that the craft will pull water from underneath it through the propeller assembly and out the top of the craft. The fin assembly controls the direction of the water flow through the craft.

In the air the craft will tilt down toward the direction that it is going. In the water, the craft will tilt up in the direction in which it is moving. The craft is very maneuverable under water. It can move in any direction.

For the best control under water, a fin assembly needs to be installed above the propeller assembly. In this way, the bottom fin assembly would be used for the major control of the craft in the air and the top assembly when in the water.

To ensure that the craft moves at the proper angle through the air and water, a vent system is operated in conjunction with the fin assembly. A better name for the fin and vent systems is the guide assembly. The vents in this assembly can close off air paths over certain areas of the craft to provide the best angle for the operating performance of the craft.

This sky and sea craft can provide the versatile application of travel in the world.

OBJECT PROPULSION FLOW THROUGH MOTOR;

In this motor configuration, the primary magnetic field is built up of a series of electro-magnets with no slot built into it. The secondary field is built up of a permanent magnet that is propelled out through the electro-magnets. As example of this is shown in figure 1022.

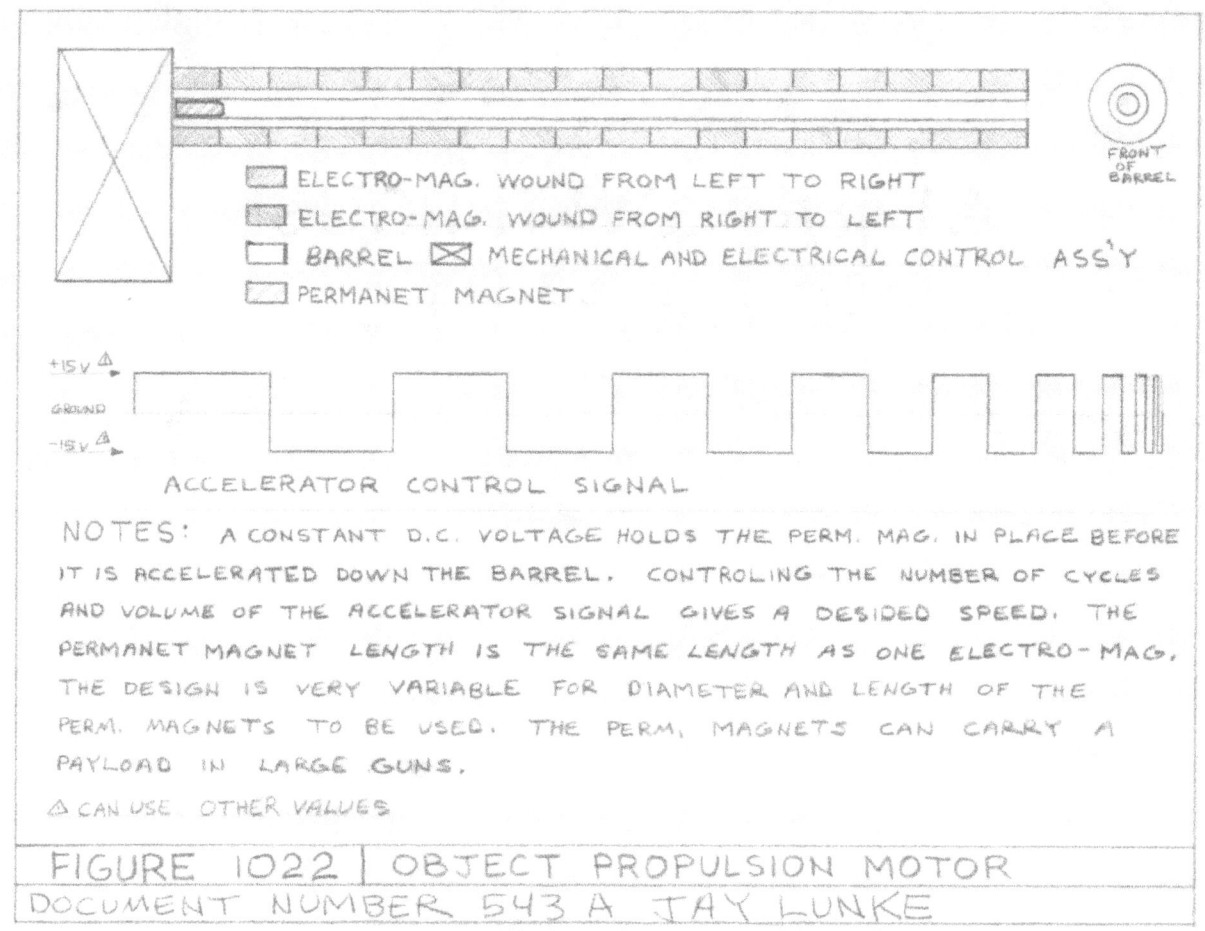

FRONT OF BARREL

☐ ELECTRO-MAG. WOUND FROM LEFT TO RIGHT
☐ ELECTRO-MAG. WOUND FROM RIGHT TO LEFT
☐ BARREL ☒ MECHANICAL AND ELECTRICAL CONTROL ASS'Y
☐ PERMANET MAGNET

+15 V △
GROUND
-15 V △

ACCELERATOR CONTROL SIGNAL

NOTES: A CONSTANT D.C. VOLTAGE HOLDS THE PERM. MAG. IN PLACE BEFORE IT IS ACCELERATED DOWN THE BARREL. CONTROLING THE NUMBER OF CYCLES AND VOLUME OF THE ACCELERATOR SIGNAL GIVES A DESIDED SPEED. THE PERMANET MAGNET LENGTH IS THE SAME LENGTH AS ONE ELECTRO-MAG. THE DESIGN IS VERY VARIABLE FOR DIAMETER AND LENGTH OF THE PERM. MAGNETS TO BE USED. THE PERM. MAGNETS CAN CARRY A PAYLOAD IN LARGE GUNS.
△ CAN USE OTHER VALUES

| FIGURE 1022 | OBJECT PROPULSION MOTOR |
| DOCUMENT NUMBER 543 A JAY LUNKE |

The permanent magnet can be solid or a shell to hold something else in it.

The electro-magnets are the same length as the permanent magnets. The electro-magnets are wound in the opposite direction as the electro-magnet next to it. This allows one power circuit to be able to operate this assembly and have the magnetic field moving flux lines in the correct direction for the devices function. When power is applied to the electro-magnet assemblies, the permanent magnet starts moving down the shaft. When the permanent magnet leaves one electro-magnet and starts entering another electromagnet, the power potential is switched to the opposite power potential. Since the permanent magnet increases its speed through each electro-magnet, the length of time the power is required to be on becomes less each time. In order to move the permanent magnet from the beginning out through the end of the barrel, an accelerator control signal is required. This signal can be produced both mechanically or electronically. There are many electronical options to produce the desired accelerator control signal.

The electronic control also lends itself to controlling the rate of speed it accelerates at for a desired performance characteristic. The object propulsion flow through motor is much safer and more accurate than other guns and cannons. The kickback is a lot less and its operation is more quiet than other guns or cannons.

OTHER FLOW THROUGH
MOTOR APPLICATIONS;

With some applications, only a partial disc is needed. In other applications, a short straight-line run can be used. Some examples of this would be a staple gun, post driver, nail driver, ext.

Almost anything you can do with another type of motor; you can do with a flow through motor. Figure 1019, 1020, and 1024 show some examples of these.

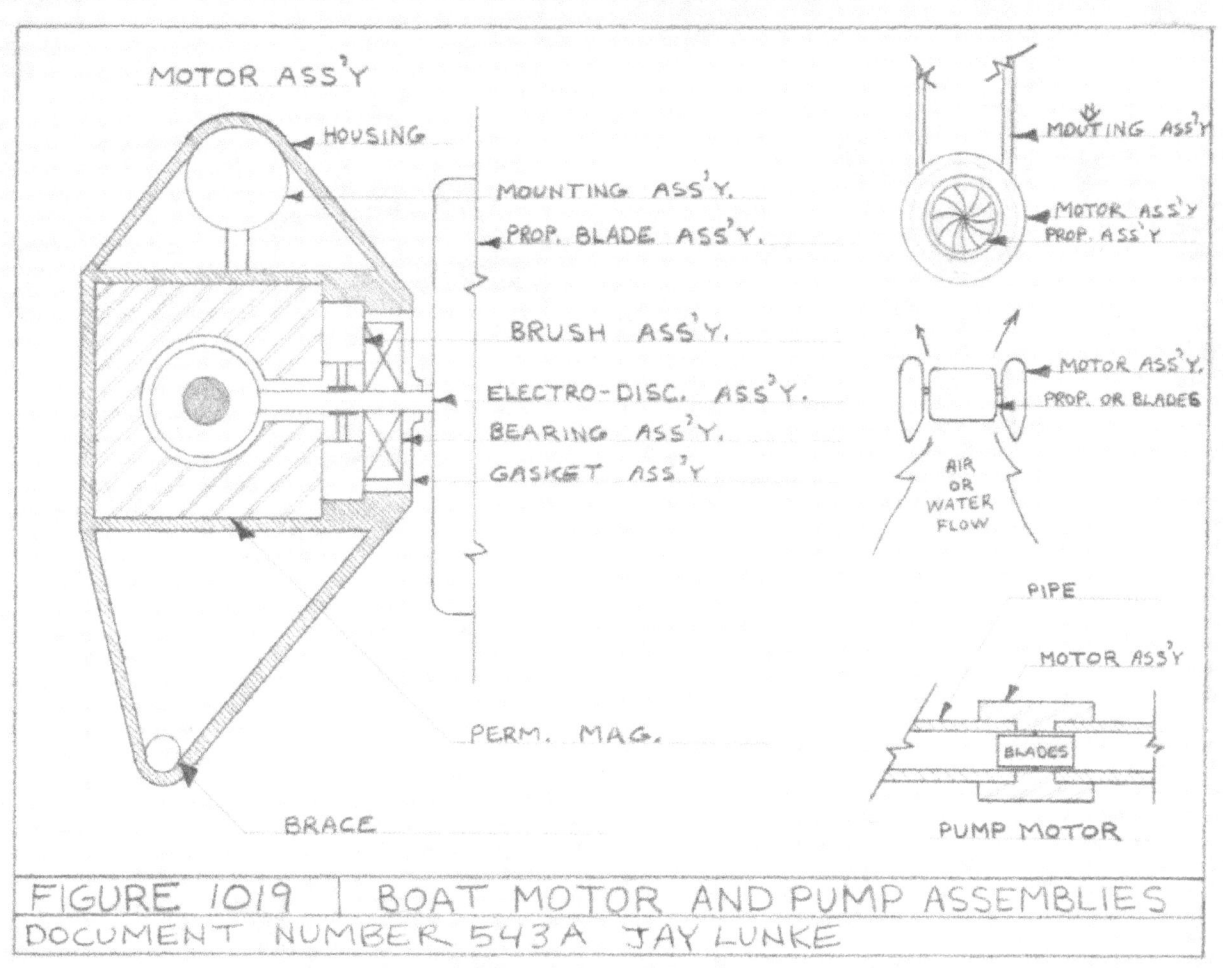

FIGURE 1019 | BOAT MOTOR AND PUMP ASSEMBLIES
DOCUMENT NUMBER 543A JAY LUNKE

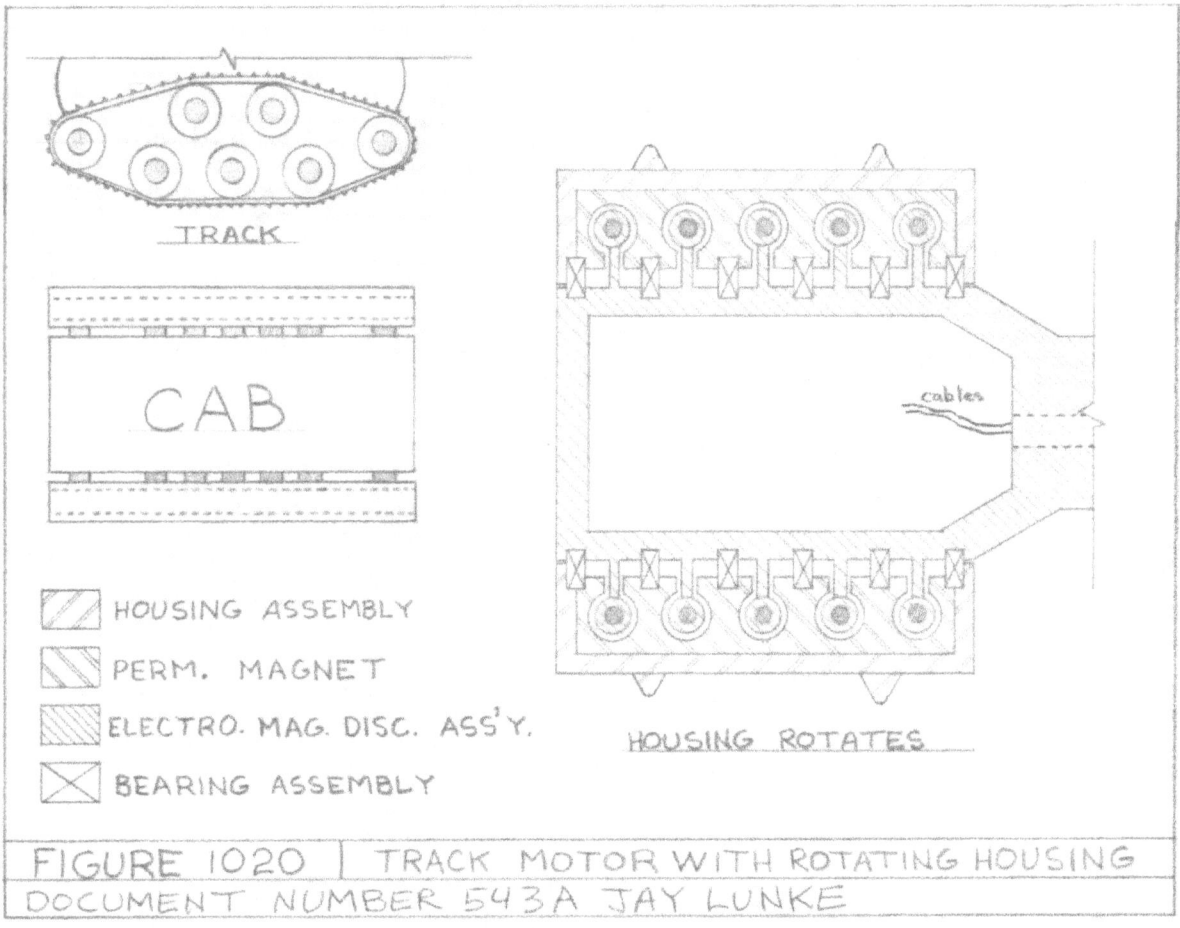

HOUSING ASSEMBLY
PERM. MAGNET
ELECTRO. MAG. DISC. ASS'Y.
BEARING ASSEMBLY

TRACK

CAB

cables

HOUSING ROTATES

FIGURE 1020 | TRACK MOTOR WITH ROTATING HOUSING
DOCUMENT NUMBER 543A JAY LUNKE

Another flow through motor application is a boat motor. One of these boat motor options is shown in figure 1019. The propeller assembly is shaped a lot different than the normal propeller assembly. Instead of being formed around a shaft, it is formed within a drum. A gasket assembly can be used to protest the critical motor parts from the water. The motor housing assembly is shaped in such a way as to pull water into the propeller assembly.

By modifying the housing assembly, this motor can be used as a pump motor. It can be mounted in a pipeline. The pump is also able to pump either direction. It can also have its pumping rate easily controlled.

In figure 10224 a submarine is built with a motor configuration with its electro-magnetic disc coming out the side of the permanent magnet ring assembly. The electro-magnet disc assembly connects to a special designed propeller assembly. Like other motor assemblies, you have the option to connect the permanent magnet assembly to the propeller assembly. The electro-magnet disc would remove the need for the brush assemblies.

The applications of this motor type being used out of the water, can include fan assemblies, swamp boats, lawn mowers, choppers, shredders, ice augers, ext.

GENERATOR AND
BREAKING SYSTEMS;

Like other motors, the flow through motor can be used in the generator mode. For a braking system, the motor can be either mechanically or electronically switched over to a generator. The generator creates a load on the application to slow it down. By connecting a variable load circuit onto the generator, you can control the magnitude of the breaking force. The energy that is produced by the breaking circuitry can be used for recharging the power source. One of these circuits are shown in the block diagram in figure 1011.

Both a variable DC voltage and a variable reverse power signal will act as a braking system. These circuits use additional energy instead or restoring it.

For an emergency braking system, it is good to use the mechanical braking system.

A variable DC voltage acts like a variable slip clutch for all the motor configurations. This characteristic can be taken advantage of in some motor applications.

In air currents you can set up a wind generator. In water currents, you can set up as electric generator.

The movement between the permanent and electro-magnet is not restricted to one direction for recharging. For example, a sky craft is tied to a dock on a windy day, As the waves rise, the propeller turns in one direction creating current flow. As the waves lower, the propeller turns in the other direction creating current flow. Both aid in recharging the power source.

Recharging techniques can only add up to applications overall operating efficiencies.

MOTOR START UP OPTIONS;

There is one dead spot per each permanent magnet for each electro-magnet. Either mechanical, electrical, or both means of correction are needed to ensure that the motor will always have positive torque on them. This will prevent the motor from having startup problems. Figure 1014 shows a mechanical method of doing this. A ring of starter permanent magnets is installed onto the disc assembly positioned close to the permanent magnet ring so that they will interact with each other.

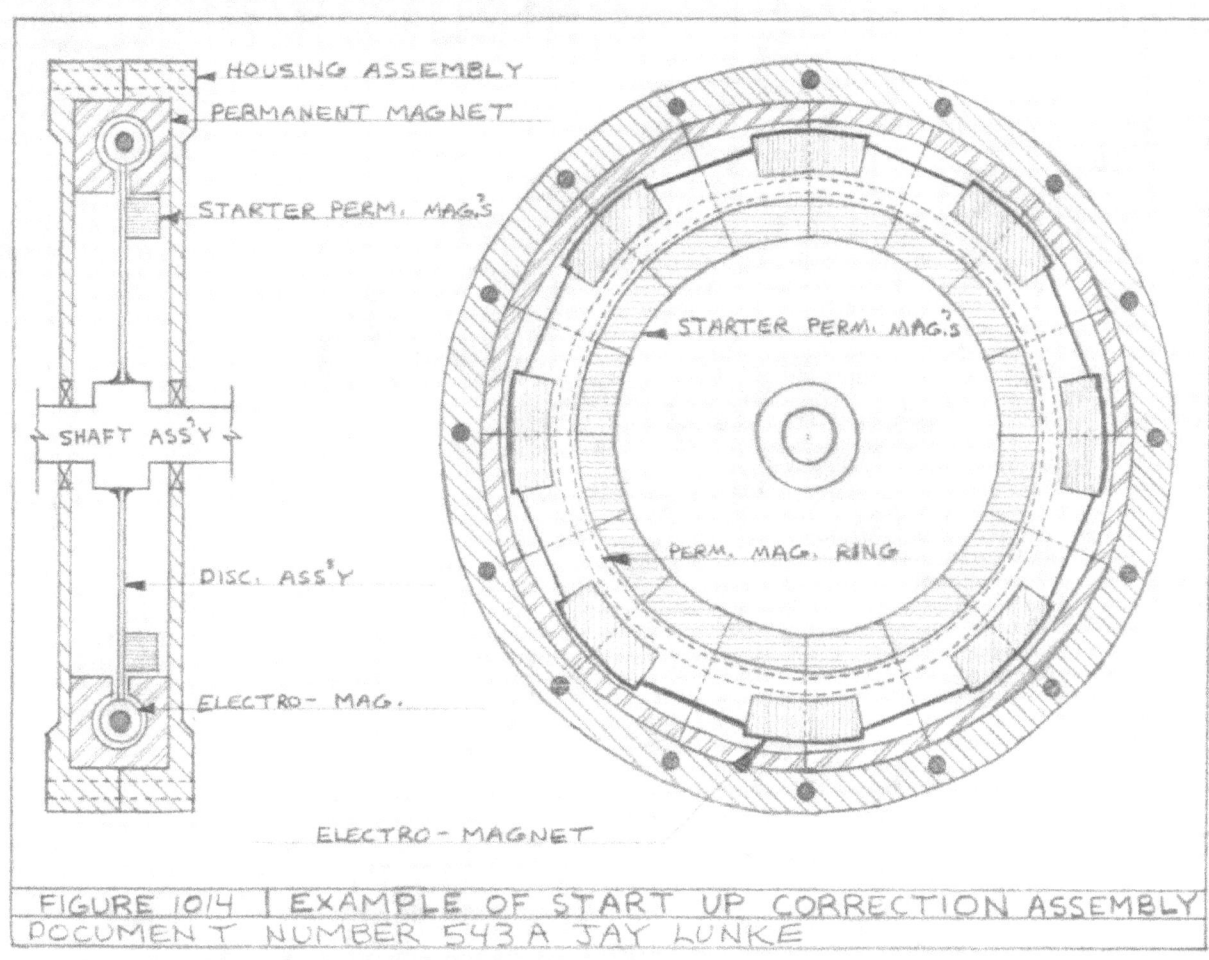

FIGURE 1014 | EXAMPLE OF START UP CORRECTION ASSEMBLY
DOCUMENT NUMBER 543 A JAY LUNKE

They will cause a positive torque, then a negative torque. You can arrange the starter magnets on the disc assembly so that when the electro-magnet is at its null point. The permanent magnet is at its peak positive torque point. The starter permanent magnets are arranged to have their maximum torque at the time the motor is at its null point. The result is to always have positive torque on the motor assembly.

Figure 1023 shows three other options to keep positive torque on the motor at all times. Either the electro-magnetic discs, or permanent magnet ring assemblies can be out of phase with each other. For a single disc assembly, the electro-magnets can be out of phase with each other. All three of these options will require at least two separate power circuits.

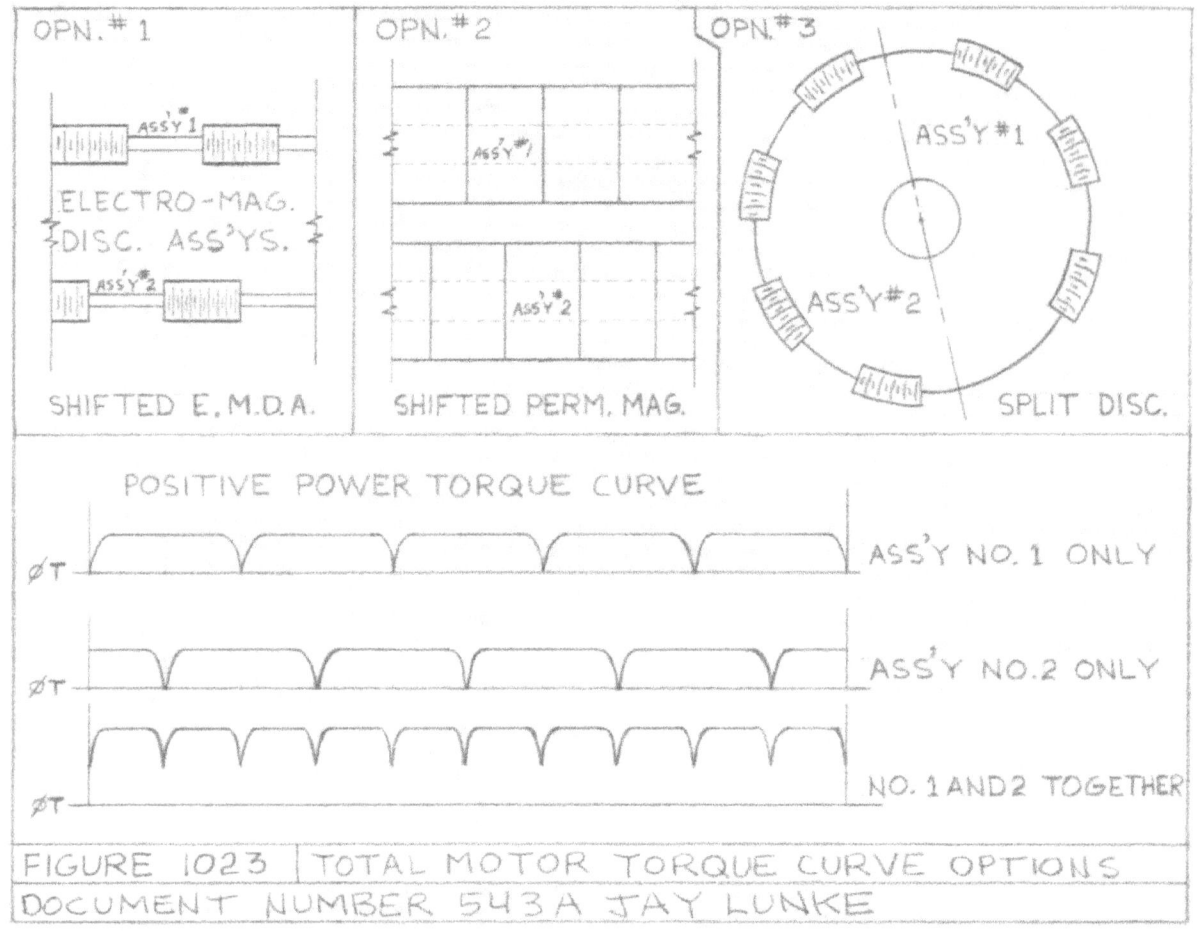

In applications where more than one motor is hooked up, the motors themselves can be operated out of phase with each other.

One additional improvement performance characteristic for having start up corrections, is that the motor will operate much smoother.

With some control circuits, the easiest way to start up the motor is to start it up like a DC motor then transfer to the other control circuits once it is running.

Final Conclusions;

The flow through motor concept provides many advantages and options over other motors. Some of them are as follows:

1. Many aspects of the motor have improved efficiency ratings. These efficiencies add up to provide a very efficient final assembly.
2. Many aspects aid to improved torque per pound characteristics of the motor.
3. The torque is adjustable and also it is much greater than that of the armature motor.

4. The large variety of flow through motor configurations make themselves versatile for many applications.
5. The motors are adaptable to a large range of power and control circuitry.
6. The motor has a long life with low maintenance requirements.
7. In some applications the motor can take advantage of the motor's gyro characteristics.
8. It includes the advantages mentioned in the other sections of this document.

The original document was completed with all of the drawings on Sept. 10, 1987. Doc.# 543A

Written by Jay Lunke

REASON FOR GIVING THE TECHNOLOGY AWAY FREE TO THE WORLD

Throughout the process of designing the motors from 1969 until today, I have prayed for wisdom from God the Father, God the Son Jesus Christ and the Holy Spirit for wisdom in designing these motors. I believed that God answered my prayers. It was not a verbal sound that I would here but more of an impression on my mind. Now I have found errors to some of my thought processes over the years that I needed to correct. So, I have always questioned whether or not they were from God. Were these mistakes a miss-understanding of the impressions. When finding the mistakes, I had an uneasy feeling about what I had written and then would investigate and find the errors and correct them. Was this uneasy feeling of errors given to me from God to correct what I had mis-understood in the first place? On one side, I do not want to say God gave me something and find out that it is totally full of holes. I guess that keeping what God could have revealed to me to myself would be a sin. God freely gave His Son Jesus Christ to die for my sins. If people kept that to themselves, then the world would be spiritually lost. Many people have shared with other people about God. It has been done freely so that they also may have the forgiveness of sins in their lives. So why shouldn't I freely give these impressions in the form of designs freely to the world as well as a thank you to God for choosing to reveal them to me in order to share with the rest of the world. To other Christians this makes sense. I know that some people have taken advantage of their position in ministry to have monetary gain, but there are a lot more people that have deprived earthly health for heavenly gain. I do not see this book as a get rich scheme on my part because it will have a very limited audience to buy it. It is a good way to publicly give this technology to the world. A person does not have to purchase this book to be able to use the technology in it to build themselves a motor. Well, some people would say that I have a motor mouth and have spoken too long in this book. I hope that you will consider reading more about Jesus Christ who is God along with the Father and the Holy Spirit by reading the Bible starting in the book of John.

May God Bless You,

Jay Lunke

www.ingramcontent.com/pod-product-compliance
Lightning Source LLC
Chambersburg PA
CBHW081558220526
45468CB00010B/2689